Handbook of Manufacturing and Supply Systems Design

Handbook of Manufacturing and Supply Systems Design

From strategy formulation to system operation

Bin Wu

Department of Industrial and Manufacturing
Systems Engineering
University of Missouri-Columbia

London and New York

First published 2002 by Taylor & Francis
11 New Fetter Lane, London EC4P 4EE

Simultaneously published in the USA and Canada
by Taylor & Francis
29 West 35th Street, New York, NY 10001

Taylor & Francis is an imprint of the Taylor & Francis Group

Publisher's Note
This book has been prepared from camera-ready copy provided by the
authors/editors.
Printed and bound in Great Britain by Biddles Ltd, Guildford and King's Lynn

British Library Cataloguing in Publication Data
A catalogue record for this book is available from the British Library

Library of Congress Cataloging in Publication Data

Wu, B. (Bin), 1957-
    Handbook of manufacturing and supply systems design / Bin Wu.
        p.cm.
    ISBN 0-415-26902-4 (hc. :alk paper)
    1.   Production Engineering. 2. Production management. 3.
        Manufacturing processes..I. Title.
    Title.
    TS176.W8 2002
    658.5–dc21

        2001044280

ISBN 0-415-26084-1

To my dear wife Sharon

and my beloved Daniel and Christopher

Contents

Preface

This is my third book in the area. With its publication, I feel that I have finally completed a trilogy on the design and operation of manufacturing and supply (MS) systems. Looking into the engineering domain of industrial and manufacturing systems, one cannot fail to notice its multidisciplinary nature. There are numerous philosophies, various approaches and techniques. Its path is paved with buzzwords. It has long been a desire of mine to be able to present a more coherent and scientific view of the area to which I have devoted my professional life.

For my part, I have always restrained myself from using too many buzzwords that happen to be the flavor of the day. This, I adhere to both in my research and in my writing. Fashions will come and go, but only sound scientific principles stand up to the test of time. That is the reason why I have attempted to follow a consistent theme throughout this trilogy of three books. The theme originates from a few words: *systems concepts, systems methods* and *systems approach*. This theme, in my opinion, provides one of the most important ways of thinking in the field. It is the basis upon which many of the so-called "philosophies" should be explained and assessed. Any sound and workable approaches must have an underlying framework that follows key systems principles. In this field, systems thinking—in terms of a set of systems concepts and prerequisites for the system's effective operation—is what provides the *necessary* conditions for any logical approach. This is my philosophy of the fundamental approach as adopted in the trilogy.

The aim of the books is to provide a comprehensive coverage of the field. Together, they serve to: (1) set systems thinking into the context of MS systems management; (2) provide a theoretical framework into which various concepts and techniques fit logically, hence illustrating *what* functions are involved, *where* they belong and *how* they can be applied; (3) present a self-contained workbook to help put the framework and techniques into practice. Accordingly, the three books cover different aspects of the subject area independently, yet their contents are complementary in a logical way.

Manufacturing Systems Design and Analysis (Wu 1994) sets systems thinking into the context of the area of manufacturing systems design. It discusses the general systems concepts and techniques, and relates these to the manufacturing domain by demonstrating the systems aspects of a manufacturing operation. In addition, it presents a structured approach for the modeling, design and evaluation of modern manufacturing systems. In essence, this book provides the systems background of the trilogy. It helps the reader to understand the structure and operation of a manufacturing organization through a systems perspective, and it shows how to use systems methods and tools to describe, analyze and design a manufacturing system in a structured way.

Manufacturing and Supply Systems Management: A Unified Framework of Systems Design and Operation (Wu 2000) provides a theoretical framework of the trilogy. Based on an extensive analysis of the available methodologies and techniques, plus results gathered through field research, it presents a unified framework of *manufacturing and supply systems management* (MSM). MSM is defined as a domain involving the activities necessary for the design, regulation and optimization of an MS system as it progresses through its life cycle. This book provides an extensive literature survey of the key topics involved in the field, and carries out an in-depth analysis of the application and future requirements of the relevant techniques. In particular, it specifies the key functional areas, outlines the contents and relationships within them, and then combines these into a closed-loop to provide the basis for an integrated management system.

Finally, this current text is all about practicality. Based on the MSM conceptual structure, this self-contained handbook guides the reader through the complete cycle of MS strategic analysis, MS system design, management of system implementation, and system operations monitoring. The structure and contents of this handbook are designed with the following in mind:

- *From the research perspective.* Many researchers involved in MS systems design and operation should find the structure of the MSM framework relevant, because it provides a logical basis for the development of consistent procedures and parameters. While researching individual methods, such a framework can help the researcher keep a systems perspective of the problem domain, and apply the resultant tools more effectively.
- *From a teaching and learning perspective.* The MSM framework will help develop a coherent view of the subject area, and aid in the understanding of how the individual concepts and techniques fit into the overall picture. The task-centered way in which the individual topics are presented will be a useful feature for lecture and tutorial preparation. The workbook itself is ideally suited for students undertaking MSM-related projects.
- *From an industrial perspective.* Industry-based professionals may utilize the workbook to plan, coordinate and execute their MSM activities in a strategically driven way. Also, the workbook is designed to assist with institutionalizing the processes dealing with system design and improvement in a company. Such an in-built ability will help a company to cope with its changing environment and demands, which is becoming increasingly crucial for the success of an MS organization.

I hope that, together, these three texts will further enhance the establishment of *manufacturing and supply systems engineering* as a scientific discipline. I can honestly say that I wish someone else had written such a trilogy, for that would have made my own life as a teacher and researcher in industrial engineering much easier!

In association with my professional activities, I have been very fortunate to receive tremendous help from a large number of people to whom I am indebted. I would like to thank a group of most highly respected colleagues: Professor R. Wild of Henley Management College, Professor J. Powell of the University of Salford, Professor D. J. Williams of Bespak Europe Inc., Professor A. K. Kochhar of Aston University, Professor D. Price of Bradford University, Professor R. J. Paul of Brunel University, UK; Professor T. J. Black of Auburn University and Professor A. Kusiak of the University of Iowa. I also wish to thank my former colleagues at Cranfield University, England, where I spent a number of very enjoyable and fruitful years.

I am particularly grateful to my colleagues in the Department of Industrial and Manufacturing Systems Engineering, University of Missouri (MU): Professors Cerry Klein, Thomas J. Crowe, Alec C. Chang, James S. Noble, Luis G. Occeña, Wooseung Jang and Jose Zayas-Castro; and Sally Schwartz and Nancy Burke. I thank them for accepting me as a colleague, for giving me the opportunity to work with a wonderful team, and for all the help that they have given as I adjust to academic life in America. I also need to thank them for their imaginative nickname for me—it is indeed great to be the *Wu at MU*.

Special thanks are due to my wife Sharon for painstakingly checking the manuscript, and for professionally converting the entire text to, alas, American English! Having studied, lived and worked in Britain for over twenty years, it took this American to force me to "agree" that the British cannot spell English properly. Of course, any errors and omissions that the reader may find in the book are entirely my own.

Finally, to Daniel and Christopher, I wish to repeat what I said in the preface of my last book: I love you guys—so very, very much!

B. Wu
Columbia-Missouri, 2001

CHAPTER ONE

A Unified Framework of Manufacturing and Supply Systems Management

1.1 INTRODUCTION

The economic and social significance of manufacturing industries has long been established: it is mainly through their activities that real wealth is created. There is little doubt that manufacturing industry will continue to play a vital role. The experiences of the manufacturing industry in the last decades of the twentieth century have provided a strong indication that the companies in the new millennium will face some new challenges.

In order to help manufacturing industries tackle the issues, a substantial amount of research has been carried out in relevant areas such as manufacturing and supply strategy analysis, and manufacturing and supply system design. Consequently, structured approaches, tools, and techniques have been developed. These have resulted in a better understanding of the processes and tasks in their individual areas. When it comes to the actual application, however, there is still a gap between theory and practice. For example, companies often still deal with their system design problems in a fire-fighting manner, due to a number of reasons identified previously. One of these appears to be a general lack of guidelines linking strategy and system design activities. Another reason appears to be the inadequate monitoring of current manufacturing system status. Without a reasonable estimate of the current status of the system in terms of its level of achievement and its position along its system life-cycle, it is difficult for the company to decide when it is necessary to initiate a new round of strategy analysis/system design activities. Also, there is a lack of integrated computer-aided tools in the area.

The issues above highlight the need for a more comprehensive framework to help companies manage their manufacturing system through the life cycles. Factories of the future will not only need manufacturing information systems to plan and control the operation of their existing manufacturing structures, but also methodologies and tools to help restructure their manufacturing and supply (MS) systems themselves. To face this challenge, the author has previously proposed a unified framework which aims to set systems thinking into the context of manufacturing and supply systems management. *Manufacturing and supply systems management* (MSM) here is defined as a functional domain that involves the major activities, such as design, implementation, operations and monitoring, etc., that are needed to regulate and optimize a manufacturing system as it progresses through its life cycle. The aim is to achieve understanding of the MSM domain, and to provide a basis for identifying a set of consistent parameters and logical procedures, so that effective mechanisms and tools can be developed to help a company's future MSM activities. Detailed discussion of this framework can be found in: *Manufacturing and Supply Systems Management: A Unified Framework of System design and Operation*, (B. Wu, Springer, 2000, London). This framework provides an MSM process reference architecture that is structured to follow the fundamental systems engineering and problem-solving principles, as well as a system reference architecture which covers the systems structure and sub-structures of an MS system. These together provide the basis for the structure of this handbook of integrated design and operation of MS systems.

This handbook has two distinctive features: it adopts a systems approach to follow through the complete cycle of MS strategic analysis, MS system design, and MS operations; and it presents MSM procedures in a task-centered and self-contained way in order to guide the user step-by-step through this cycle. Together, the MSM framework and its task-centered workbook help set systems practice into the context of MS system design and operation. They present an integrated MS systems management framework, logically incorporating the principles and key techniques from a number of relevant areas, including:

- systems concepts and systems engineering,
- systems structure and systems perspective of MS operations,
- strategic planning and objectives formulation,
- system design methodology and techniques,
- project and change management, and
- system performance monitoring.

Following the key principles of systems theory and techniques, the remainder of this chapter provides an overview of the conceptual structure of the MSM framework which identifies the main functional areas, specifies their generic functionality and contents, and logically integrates them into a closed-loop to provide the basis for effective systems management. The task-centered workbook will be presented in the subsequent chapters of the book. Issues related to the framework's institutionalization within an MS organization will also be discussed.

1.2 BACKGROUND AND KEY ISSUES

The last two decades of the twentieth century have seen a new approach to manufacturing. The new demands from the customers and the market have resulted in a reduction in product life-cycles, and hence the need to reduce the time-to-market period for new product development. In addition, it is no longer possible to merely exist and compete at a local level. Competition is seen to exist on a global scale, with world class standards being set in many areas.

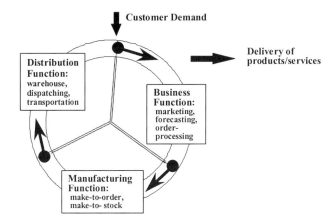

Figure 1.1 Three main functions contributing to MS performance

For many decades, manufacturing and other system functions, such as marketing and distribution, have been treated as separate activities in a manufacturing organization. They may no longer be treated in such a way: in today's global setting, the success of a manufacturing organization can only be achieved with the optimization of the manufacturing and other functions in logical association with one another.

For instance, the importance of transportation/distribution within the manufacturing domain is highlighted in Figure 1.1. This shows that, from a customer's point of view, there are three main functions contributing to a company's delivery performance. This makes it quite clear why companies are increasingly using their supply chains as competitive weapons. Hence, logistics and manufacturing are linked together in an organization's overall manufacturing and supply operation, frequently making the structure of the organization a distributed one involving manufacturing/supply units at different sites and geographical locations (Figure 1.2).

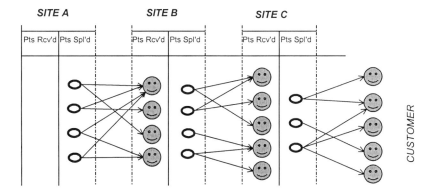

Figure 1.2 Structure of a distributed MS operation

Optimization of the complete manufacturing and supply cycle has increasingly become an essential determinant to gaining a competitive advantage. However, current techniques of manufacturing strategy formulation and system design seem to have concentrated mainly on the issues related to manufacturing activities alone, without much consideration being directed to their subsequent operations. It is evident that many companies have found this restricting, and have begun to ask for ways to consider these relevant activities and treat them as an integral part of the complete cycle. For many manufacturing companies, reaction to market and business conditions suggests the requirement for a step change followed by continuous improvement. This in itself is likely to be continuous, needing steps or sprints in performance to be achieved periodically, with incremental changes occurring in between. Consequently, MS system design (MSD) projects are being carried out much more frequently than before. Similar to what is known as a *product life cycle*, a manufacturing and supply system also possesses a life cycle, going through a series of stages as shown in Figure 1.3. As shown, greenfield type system design projects are required when a completely new system is introduced, designed, and implemented to satisfy a new set of manufacturing requirements. The subsequent system design activities, brought about by continuous improvement initiatives and projects responding to new market requirements, can be referred to as *continuous improvement* or *brownfield* type projects. In both cases, it is generally necessary to carry out a redesign project, requiring the utilization of existing resources, and being subject to constraints related to the existing system. This concept of *MS system life cycle* provides an insight into the reason why today's manufacturing organizations have to become more lean and agile.

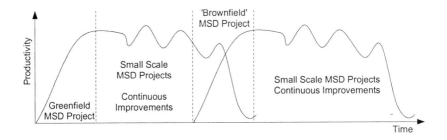

Figure 1.3 MS system life cycle

In reality, every case is different. Companies commence system design projects from different perspectives. Not only are the markets different in many cases, but each enterprise possesses a unique history, a unique organizational culture and state, and a specific strategic direction. Other factors, such as the combination of time, resources, and financial constraints are also specific to individual companies. Therefore, a design process should be adapted to suit the particular case, requiring an appropriate means of guiding the organization through the relevant design tasks. Such an approach would need to consider the entire design process from the setting of objectives to the detailed design stages and the specification of implementation activities. Based on an extensive analysis of the relevant literature and results from practical cases, a number of separate, yet interdependent, key issues in the area can be summarized as follows:

- *System design methodologies.* There are several problems regarding the use of manufacturing system design methodologies: (1) Awareness—the actual application of methodologies in practice; (2) Planning—the requirement to encourage coherence in the tasks undertaken from project initiation through to implementation; (3) Documentation—the recording, manipulation and retrieval of design data such as design notes, assumptions made and their justifications, etc.; (4) Implementation—failures in system design projects are often related to inadequate organizational and operational planning and/or faulty execution of the implementation process. The primary areas of concern include the lead-times of projects, decision making, and the insufficient coordination of tasks.
- *Manufacturing strategy analysis/formulation.* A few relevant issues in this area are: (1) *Manufacturing strategy formulation*—this covers the strategy content and process. There is substantial agreement concerning the decision categories or manufacturing policy areas to be addressed within a manufacturing strategy; (2) *Interdependency between strategic policy areas*—the decisions made for the manufacturing policy areas are interdependent. The policies and ensuing system design activities can logically link functions to strategy, or can involve more complex multiple links between functions; (3) *Audit approaches*—these allow a systematic involvement of key personnel, and allow both data and judgments to be recorded and revisited.
- *Strategy/system design interface.* Strategy formulation can highlight both strategic improvements and operational improvements that can be achieved through system design activities. The planning and formulation of a design project should be assisted by strategic plans, by identifying *cause and effect* relationships between strategy and operations. The plans derived from the manufacturing strategy should concern the definition of implementation requirements for the manufacturing policies, the definition of the basic manufacturing systems and procedures, the definition of manufacturing controls, the selection of operations critical to manufacturing success, and the definition and formulation of performance measures and review procedures. However, the process of strategy formulation and its subsequent derivation into the specification of action plans is currently considered to be mainly creative. A significant feature resulting from this fact is that the action plans are often not sufficiently detailed to aid implementation. Since strategy development is an iterative process, it should be useful to consider iterations across the strategy/system interface throughout the system design project, though particularly at the early stages. These iterations may also feed back to the top level corporate and business strategies where necessary. Strategy/system design interface can be viewed as being a complementary task to that of strategy planning and specifies how the strategy is to be executed, the resources required and the performance measures to be applied. It can therefore be considered to occupy the phase of the interface that concerns the development of action plans. In a tactical sense, these plans represent individual system design projects.
- *Systems status monitoring.* This area raises issues about strategy/system implementation, and how to judge the success/effectiveness of a project. A problem has been observed with respect to knowing where to start a system

design project. The reason appears to be the lack of an online monitor of current system status within the MSM context. It must be realized that, in order to effectively support a strategy, the development and implementation of the necessary system and operations are a continuous process. Once a new system is implemented, its performance needs to be regularly monitored to assess its fitness-for-purpose, so that the original strategic goals are achieved.

1.3 SYSTEMS APPROACH TO MS MANAGEMENT

In order to deal with the complexity involved, the systems approach to the design and operation of modern MS system, as presented in one of the author's previous books on the subject (Wu, 1994, *Manufacturing System design and Anal*ysis, 2nd Edition, Chapman and Hall, London) has become more relevant than ever. The structure of the proposed MSM framework closely follows the systems principles and the prerequisite conditions for effective system construction and operation. It essentially supports a structured mechanism for the provision and execution of relevant MSM methodologies, and the communication of system designs.

1.3.1 Key Systems Requirements

Amongst the various concepts as presented in the above mentioned text, of particular interest are a prototype system model and its set of conditions necessary for the effective operation and control of manufacturing organizations. As far as the development of the MSM framework is concerned, the following are especially relevant: (1) *Coherent organizational and operational strategies.* The objectives adopted at various levels of the system must be in line with the overall business aims. Therefore, regardless of the type of system design projects concerned, their activities should be strategically driven so that they are carried out following a coherent frame of objectives to guarantee the system's fitness-for-purpose; (2) *Adequate system structure.* In order to achieve the first goal, a hierarchy of closed-loop control mechanisms must be implemented which corresponds to the hierarchy of manufacturing and supply functions. Hence, three fundamental system functions must be properly designed and implemented at each level along the hierarchy— objective setting, operational and performance monitoring; (3) *Adequate measurement of the processes.* To facilitate an effective control, it is necessary to be able to measure relevant process parameters in an adequate manner, highlighting the need for the current system performance to be adequately estimated for the subsequent decision-making within the MSM loop; (4) *Awareness of environmental influences.* Sufficient consideration must continuously be given to environmental factors, including changes in customer requirements, technological development, competitors/partners' level of achievement, and changes in government regulations and economical climate. If one relates these well-proven systems principles to the area of MS management, it becomes apparent that a few key elements should be logically incorporated into an overall framework, so as to provide a logical and practical MSM management approach.

1.3.2 Overview of MSM Framework Structure

As shown in Figure 1.4, the MSM framework should consist of three main functional areas: manufacturing and supply strategy analysis (MSA), manufacturing and supply system design (MSD), and manufacturing and supply operations management (MSO). Generally speaking, the nature of MSA approaches can be summarized as a method of helping a company analyze its products, market, and operations to identify areas of concern, and then setting objectives for improvement. However, the implementation of strategic initiatives will rely on the management of change through MSD projects. The general aim of an MSD project can therefore be defined as the determination of the best structure of a manufacturing and supply system in order to provide the capability needed to support strategic objectives. This must be achieved within the resource and other constraints. An MSD procedure is usually based upon a general model of a problem solving cycle, as exemplified by the MSD methodology outlined previously by the author. In addition, the complete MSM cycle should also include the aspects of manufacturing and supply to plan, monitor, and control the production processes once the system is implemented and in operation. Therefore, the MSO area largely reflects the planning and control activities normally associated with an manufacturing resource planning (MRP)/enterprise resource planning (ERP).

Figure 1.4 Overall functional structure of a unified MSM framework

The systems thinking in the management of manufacturing and supply requires the development of a set of coherent strategic objectives and goals. The message bears repetition: a hierarchy of compatible system structures should support this hierarchy of objectives. Failure to deploy such an approach will tend to produce solutions/systems that may be technically good but not necessarily good for the business as a whole, due to a lack of context and coherence. In close relations to the MSA function, therefore, a core area involving costing, quality assurance and performance measurement is specified. Its role is to provide a coherent means of establishing goals and objectives, and evaluating the output from various functions in a way that is consistent with the overall strategic aims.

The overlap between these main areas identifies three additional MSM functions: MSA/MSD interfacing, MS system implementation and MS system status monitoring. One particular feature of this framework is the inclusion of this system status monitoring domain. Its function is to regularly monitor the system's performance against the original strategic goals. Modification of the system structure, operational procedures, and even the original strategic contents can subsequently be necessary. Accordingly, the purpose of this system status monitor is to assess the system's current performance, identify its status along the life cycle, and to trigger appropriate MSA/MSD projects when necessary.

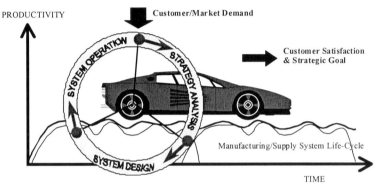

Figure 1.5 MSM as the driving-wheel of a manufacturing/supply organization

Therefore, from a theoretical point of view, the proposed framework reflects the fact that a systems approach should be adapted to the design, implementation, and management of manufacturing and supply systems. From a practical point of view, on the other hand, such a framework aims to provide a means of coordinating and supporting the relevant MSD tasks, monitoring and operating the resultant MS system in a strategically driven way. As a result, the shape, size and dynamic characteristics of the system are fit for the purpose, capable of coping with the demands put upon it, and able to achieve its strategic goal along the rather bumpy route of its life cycle, as shown in Figure 1.5.

1.3.3 Overview of MS System Structure

Just like the engineering design of a product, the complete specification of an MS operation will have to include a number of documents and drawings, each of which provides information about the structure or function of a specific part of the system. Therefore, in addition to specifying the structure and sequence of the analysis processes themselves, the framework also provides a means of describing the resultant system. That is, it provides a design process reference architecture, as well as a modeling reference architecture, which covers the MS systems structures, and sub-structures. At each stage, a number of MS sub-systems can be addressed. Three principal MS functional areas can be addressed through MSM activities within this framework:

- The *physical (or manufacturing/supply process) architecture* represents the 'hard' elements of the manufacturing and supply systems, including the machines, transportation and storage equipment and the other facilities required to support the manufacturing and supply process. This also describes the flow of materials through the system.
- The *human and organizational architecture* represents the organizational structure and the interactions of the employees within the manufacturing and supply system, including their roles, responsibilities, and tasks.
- The *information and control architecture* represents the planning and control functions of the manufacturing and supply system and the processes involved in decision making. This also describes the flow of data and information in all its formats, whether paper or computer based, throughout the system.

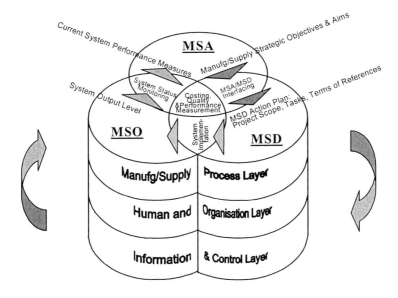

Figure 1.6 Overall functional structure and flow of a unified MSM framework

Consequently, the complete functional areas and their logical sequence are as shown in Figure 1.6. This figure illustrates the continuous processes through the complete MSM cycle that need to be considered within the three layers of a manufacturing and supply system. The overlapping domains of these three architectures provide three further design concepts: the system structure, system decisions, and system functions, which are outlined in Figure 1.7. Hence, the functionality of an MS system is provided through the combination of physical MS facilities to carry out the transformation processes; the organization of the physical facilities and personnel to provide the system structure; and the information structure to define how and what the system should produce. By using these architectures and concepts, a direction for system design and modeling can be formulated. Progressing from the center, the requirements with respect to the system concepts can be specified in a holistic manner and the individual architecture and sub-systems' requirements can be defined.

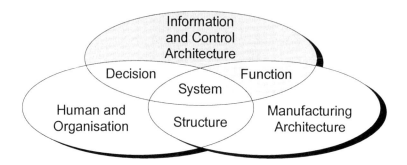

Figure 1.7 A conceptual MS systems architecture

The above is further refined by following the generic MSD methodology previously developed by the author (Wu 1994), consisting of four stages: project initiation, requirements specification, conceptual modeling and detailed design. The project initiation stage provides the terms of reference for the particular MSD project. The requirements definition stage provides a specification for the MS system. The conceptual modeling stage generates a number of alternative configurations for feasibility assessment. Finally, the detailed design stage provides the opportunity to render an in-depth specification of the chosen conceptual configuration, as shown in Figure 1.8.

Existing System Analysis	Manufacturing Strategy	Manufacturing Criteria	Project Initiation
Terms of Reference and Project Approval			
System Function	System Structure	System Decisions	Requirements Specification
System Requirements and System Approval			
Manufacturing	Information and Control	Human and Organisation	Conceptual Design
Feasibility Study			
Process / Transport / Support / Planning / Control / Organisation / Human / Facilities			Detailed Design
Implementation Requirements			

Figure 1.8 Overall MSA/MSD tasks and reference structure

1.4 THE MAIN MSM FUNCTIONAL AREAS

These are the areas where a substantial amount of research has already been carried out. Consequently, structured approaches, tools, and techniques are available to help with the tasks involved.

1.4.1 MS Strategic Analysis

The purpose of the first functional area is to help develop and capture a company's future MS strategy (Figure 1.9). Long-term success requires a company to continually seek new ways of increasing its overall efficiency, and of differentiating itself from competitors so as to enhance its particular competitiveness. To create such a strategic approach, a company must develop a plan for identifying and building the capabilities that will enable it to do certain things better than its competitors can.

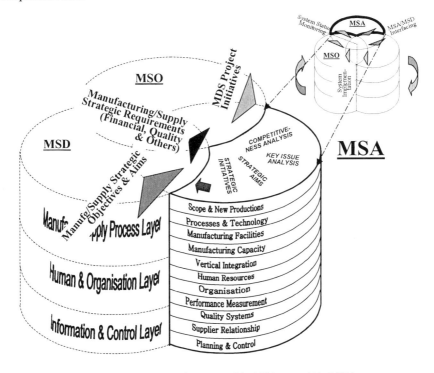

Figure 1.9 Processes and contents of the MSA area within MSM

As far as research in MS strategy is concerned, a general model is usually followed which broadly divides MS strategy into two related domains of MS strategic process and strategic content. *Process* refers to the procedure of formulating and implementing strategy and *content* refers to the choices, plans, and actions that make up a strategic direction. Several approaches to the formulation of

MS strategy have been published in the literature. An analysis of these has indicated that, with respect to strategic content variables, there is a significant degree of agreement amongst the current approaches. This has enabled a generic MS policy model to be developed, as shown in Figure 1.9. This model consists of eleven policy areas. Each policy area has been defined with respect to its decisions, sub-decisions, options, parameters, and influences.

The underlying logic of a typical approach to MS strategy formulation follows that of the generic problem-solving model. That is, it may be best illustrated by the situation where one wishes to travel geographically from location **A** to **B**, and plans the journey by asking questions such as: what is our destination? where are we now? what are the possible routes and means of transport? which route is best?

Similarly, to accomplish the best system changes in MS, both the starting point and the desired state should be known. It is then necessary to understand how the current system can best achieve the current or future requirements. This can be accomplished by identifying the reasons for the problems and the most effective means of filling the gaps. It essentially consists of three important task frames: *existing system analysis*, *MS strategy* and *MS criteria definition*. Each of the frames represents an independent and complimentary set of tasks. Although these particular tasks do not contribute towards the specification of an MS architecture or sub-architecture, they can be viewed in a global systems-wide perspective. The combination of these task frames provides the terms of reference for an MSD project. Additional concerns, such as the organizational and business strategies, can also provide an indication of the future MS system:

- *Existing system analysis*. The existing system analysis provides an initial pre-MSD diagnostic of the MS system and its operating environment. In cases where an MS strategy formulation exercise has been undertaken, or where significant operational problems have been highlighted and investigated, such an analysis will already have been completed. If this is the case, the pertinent information derived from the exercise should be recorded. The analysis is not rigorously detailed at this stage of the project, but it does provide an indication of the performance of the MS system. It also serves to highlight any problems and problem areas. The principal tasks in this frame include: (1) *Product grouping.* This serves as a means of aggregating the many separate items likely to be produced within the MS function into sensible product families that share common attributes or properties. A number of analytical methods are available to assist in the process, such as the use of the product process life-cycle matrix and the criteria matrices based on the competing characteristics of product families. Once the basic product families have been identified, it is useful to be able to rank these in a relative sense with respect to their importance to the business; (2) *SWOT analysis.* The strengths, weaknesses, opportunities and threats for each Product group should be identified and recorded; (3) *Performance analysis.* The performance analysis with respect to the customer requirements should be carried out in a disaggregated manner, typically based upon the Product groups derived in the earlier design task; (4) *Problem identification.* A separate design task, problem identification is used to highlight problems prevalent amongst the Product groups and to attempt to locate the root causes of these problems.

- *MS strategy.* The strategy analysis aims to capture the relevant information contained within the enterprise's MS strategy. The key inputs to this task frame are therefore the MS strategy document, the operating plan document, and the action plan document (if they exist). The information contained in these documents can be collected in a structured manner together with ancillary information that may have been generated during the formulation of these documents. It is expected that the results are generated either within the existing system analysis stage or within one of the MS strategy formulation approaches. Based on the results from the survey on the current approaches, Figure 1.9 illustrates the generic MS strategies frame that provides a basis for strategy capture and the subsequent selection of MSD activities. Additional information is provided to assist the users in identifying problems within their system and acts as a checklist for each individual competitive criteria and MS policy area. These present typical problems prevalent within the policy areas and indicate likely effects on the competitive criteria of the MS function.
- *MS criteria.* The MS criteria essentially provide an indication of the customer requirements with respect to the MS system in strategic terms. They are mainly derived from the business and MS strategies. The criteria are grouped into: (1) *System purpose.* This defines the rationale and aims of the MS system, with respect to its role in the organization, including the direction in which it is heading and its functionality. Hence, this criterion includes concepts such as the product range, customer demand, volume manufactured, and the core processes of the MS system; (2) *System performance.* This is concerned with the quantitative measures of the system with respect to its competitive performance. Competitive criteria include product lead-times, customer lead-times, delivery dependability, quality levels and scrap rates, etc.; (3) *System characteristics.* These are the non-quantifiable criteria of the system and cover a qualitative assessment of the systems operations (such as the degrees of simplification, automation, and integration, and the degree of system flexibility); (4) *System costs.* These relate to the financial aspects of the MS system. They include targets for fixed-assets investment costs, materials, and inventory costs, and operational costs.
- *Consolidation.* This brings together all the design and strategy information captured, created and generated previously. The information will be presented to the designers/managers. They then verify the consistency and check: (1) *Readiness for change.* This is an indication of the organization's readiness for change, in terms of implementing a new MS strategy, reorganizing its MS operations and executing an MSD project. A series of questionnaires and worksheets are presented to assist in the assessment of the organization's preparedness; (2) *Terms of reference.* This provides MSD-specific aims and constraints which summarize the project scope, project constraints, system constraints and project objectives. The project scope is classified into six categories: project initiators, product-system type, project focus, project type, desired solution and project level. The project constraints and the system constraints are each classified into four categories: time constraints, resource constraints, human resource constraints, and financial constraints. Finally, the project objectives are classified into four categories: financial, quality, organizational, and operational.

1.4.2 MSA/MSD Interfacing

Figure 1.3 has clearly indicated that, in reality, every MSD project is distinctive and has different scope, concern, and strategic objectives. Therefore, it is important that a company should be able to identify the relevant options and related MSD tasks so that their MSD actions address the key issues to achieve the required improvement. A generic MSA/MSD interface has been developed within the MSM framework to enable manufacturing companies to make more informed decisions in this regard (Figure 1.10). Using MS strategic initiatives as the principal input, this interface aims to assist in the association between MS strategy concerns and necessary system design actions. The first stage is concerned with MS requirement specification—the definition of the system with respect to its function, structure, and decisions (see Table 1.1):

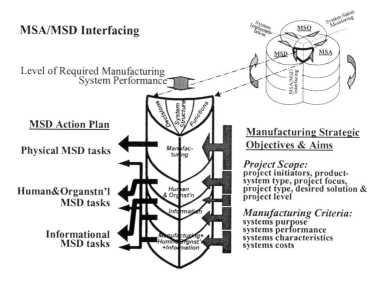

Figure 1.10 MS Requirement specification

- *System function.* This aims to provide a more detailed definition of the purpose of the MS system, as previously outlined in the system purpose category of the MS criteria. The task frame builds on the information supplied by the previous stage, both explicit in terms of the products to be manufactured and supplied, and implicit with respect to data applied within the MS strategy formulation and analysis. Hence, it aims to define the required function of the system with respect to current and future products, and the associated processes, both in-house and subcontracted.
- *System structure.* The systems structure task frame specifies the overall structure of the MS system. It covers the definition of the functional grouping of the system and includes system decisions such as capacity planning. Hence, it aims to define the required structure of the MS system with respect to the process

organization and grouping of MS functions and the degree of modularization and integration within the system.

- *System decision.* The decision task frame identifies the necessary requirements for the information and control systems of the MS system. It also specifies how these interrelate with the physical and organizational sub-systems. Therefore, it aims to define the required decision making structure of the MS system with respect to the decision and control processes, and the degree of integration within the system.

- *Consolidation—system requirements.* The final section of the requirements phase of the MSD process is the consolidation stage. This brings together all the design requirements and strategy information captured, created and generated within the three task frames. The results of this stage should include a definition of the system boundaries and those being addressed within the MSD project; a definition of the systems architectures being addressed; a model of the requirements for the system; and a project objectives definition. The information is presented to the designers for evaluation and verification of consistency and completeness. Finally, an MS systems requirements report is generated. This ensures the continual communication of results within the organization and provides a high-level approval and checking mechanism with respect to the consistency of the initial system specification with the overall business and MS strategy policies and goals.

Table 1.1 Requirements specification

	MSD Task	**Description**
System Function	Product Analysis	Specification of requirements of new and existing products
	Part Analysis	Specification of requirements of new and existing parts
	Process Analysis	Specification of processes and process technologies
	Make vs. Buy	Analysis of processes for in-house or subcontract
System Structure	Functional Grouping	Specification of functional groups (process or product)
	Capacity-Demand	Specification of capacity required for each group
	Structural Layout	Specification of MS organization and structure
	Integration-Modularization	Specification of degree of modularization and integration and identification of individual modules
System Decision	Information Functions	Specification of information functions
	Decision Variables	Specification of level of decision making, level of control, decision-making hierarchy

Following the above, the MSA/MSD linking process is supported by a series of generic action plans. Each of these plans is associated with a set of MSD tasks derived from the MSD functional area. In fact, the MSD functional area has been specified in such a way so that, to a certain degree, it corresponds to the generic MS policy areas. Figure 1.11 illustrates the MS policy areas and associates them with the MS sub-systems that are typically addressed in an MSD project. These relationships between the policy areas and MSD task frames are relatively simplistic, particularly when the multiple-interdependencies of MS strategy and MS systems are considered, but they do provide a logical indication of the dominant sub-systems and task frames that initially need to be addressed in the design process. In reality, however, due to the interdependencies amongst policy areas themselves, and between policy areas and design tasks, a top-down approach for linking strategy to the design process can only be established in several stages.

The initial strategic objectives generally provide a qualitative and/or quantitative indication of future directions for the organization, based on the differences between what the market requires from the company and the actual performance of the company's MS system. In addition, the MS criteria defined through the MSA process relates MS strategy to MS system by defining the system purpose, system performance, system characteristics, and system cost structure. Following these, a number of MSA/MSD link-tables are provided, indicating cause-effects relationships. They form an MSA/MSD linking chain through the following steps:

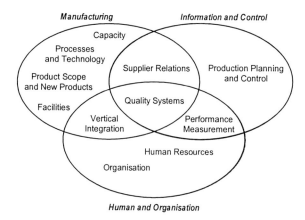

Figure 1.11 Conceptual relationships between MSA policy areas and MS sub-systems

- *Strategic decisions—MSD tasks.* This provides an indication of the possible relationships between each of the sub-decisions, categorized under the strategic decisions of each of the eleven MS policy areas and the approximately seventy-five MSD tasks of the MSD task framework. There are currently over two hundred separate sub-decisions grouped under fifty-five decisions within the eleven policy areas. When the table is analyzed, it can be seen that the mapping functions linking policy areas to sub-systems, as illustrated in Figure 1.12, though simplistic in nature, correspond sufficiently with the more detailed level of abstraction. As well as supplying information for the selection of the relevant MSD tasks, this table also provides links from the design tasks back to the strategic policy areas, decisions, and sub-decisions. Hence, when a design task is being performed—whether at the task refinement or execution stage—the user can refer back to the associated strategy decisions for guidance and check its consistency on a global level.
- *Generic action plans—MSD tasks.* This provides an indication of the possible relationships between each of the generic action plans and the MSD tasks of the

MSD task framework. Altogether, eighty-eight generic action plans represent an aggregation of those identified in the literature and those observed in industrial practice from case studies. They provide a broad cross-section of the types of MSD projects and action plans likely to be required, from complete MSD projects to continuous improvement programs.

- *Project terms of reference—MSD tasks.* This provides an indication of the general relationships between each of the project terms of reference and the MSD tasks of the MSD task framework. Just as the strategy-design task table is applied, the terms of reference-MSD task table can be used to refer back to the relevant project terms of reference during design task execution.

The linking tables used for the MSA/MSD interfacing can be edited by the users to match the specific strategic requirements of their enterprise. New entries can be added, relationships can be changed, and their respective weightings altered. Again, a workbook approach has been followed, outlining steps to guide the user through the process and presenting the user with logical options.

Policy Area	Decisions	Sub-decisions	Product analysis	Part analysis	Process analysis	Functional make vs buy	Functional grouping	Capacity - demand	Structural layout	Integration - modularisation	Information functions	Decision variables	Process planning	Part grouping	Make vs Buy	....
C	Total capacity	Demand pitch						x			x	x				...
A		Floor Space						x								
P		Plant			x	x		x							x	
A		Equipment			x	x		x							x	
C		Labour						x								
I	Variation Satisfaction	Cyclical	x	x	x	x		x							x	
T		Long Term Trends				x		x								
Y		Demand Highs				x		x							x	
		Demand Lows				x		x							x	
		Degree of flexibility	x	x	x	x	x	x	x	x			x	x	x	
	Expansion Methods	How						x	x						x	
		Size of increment						x	x						x	
	Contraction Methods	How						x	x						x	
		Size of decrement						x	x						x	
	Timing															
	Bottlenecks															
	Demand forecasting	How monitor									x	x				
		How forecast									x	x				
		Cap. change signal									x	x				...
F	Number							x								...
A	Specification	Size						x								
C		Capability	x	x	x	x										

Figure 1.12 Sample strategic decision/MSD task relationships (partial listing)

1.4.3 Task-Centered MS System Design

An MSD methodology usually follows a structured cycle that involves a number of typical stages such as "formulation of objectives" followed by "conceptual design" and "detailed design". More detailed MSD tasks can be specified at each of these stages. A literature survey has been carried out on these methodologies and other relevant approaches to the design of MS systems. In addition to these methodologies, specific techniques for system design have also been reported, although these aim to deal with particular MS sub-systems, and, often, particular aspects of sub-systems. From this analysis, it is possible to identify a set of generic MSD tasks that are carried out along a process of analysis and evaluation, as shown in Figure 1.13.

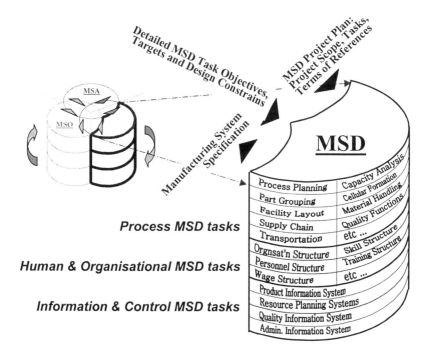

Figure 1.13 Contents of the MSD area within MSM

The MSM structure provides an effective basis for the clarification of its functional domain. The functional area is divided into individual cells, each of which represents a particular module. The modules' specific contents (functionality, relevant techniques, parameters, values, relationships, etc.) may be specified in detail, if required, as illustrated in Figure 1.14. Within such a task frame, which can be considered to represent a self-contained package of work, a design task collection exists that addresses a specific sub-system at a particular stage in the design cycle. Hence, it is within these generic frames that sub-problems are solved and a design concept developed. It is through selecting appropriate task frames and design tasks, and through customizing their contents, that a specific MSD project can be defined. Thus, not only can the design of a modular system be created with respect to production units, manufacturing cells and workstations, but the actual design process itself can also be modularized according to the sub-systems addressed and design tasks chosen and executed.

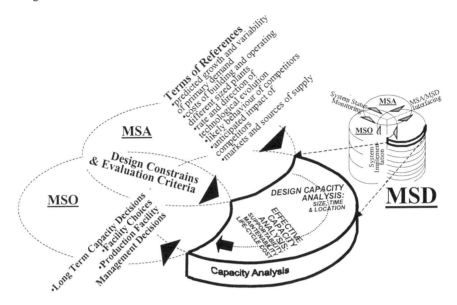

Figure 1.14 MSD task example—capacity analysis

Conceptual design

Within the conceptual design stage, a number of alternative MSD options can be generated and assessed based upon the requirements, terms of reference, and strategy developed previously. The conceptual design stage is based around the three sub-architectures. Its aim is to identify a number of approaches that may fulfill the system's requirements. As such, it needs to take into account the existing system's structure and functionality, as well as any constraints imposed by the existing system. It consists of the MSD tasks shown in Table 1.2:

- *MS processes.* The purpose of this task frame is to specify the physical entities of the manufacturing and supply system at a conceptual level of detail. Hence, it is concerned with the physical processes, services, facilities and support required, as well as the overall capacity and conceptual layout of the system.
- *Human and organization.* The purpose of this task frame is to specify the organizational entities of the MS system at a conceptual level of detail. Therefore, it is concerned with the structure, culture, and state of the organization supporting the physical and information systems, and the general operating policies of the organization. Quality issues are also addressed within this task frame.

Table 1.2 Manufacturing MSD task documents

	MSD Task	Description
Manufacturing and Supply Processes	Process Planning	Verification and specification of process plans for part
	Part Grouping	Specification of part groups according to a variety of
	Make vs. Buy	Make versus buy analysis (parts)
	Cell Formation	Specification of cells according to a variety of criteria
	Conceptual Layout	Conceptual modeling of factory layout
	Conceptual Capacity	Specification of required capacity of individual cells
	Space Determination	Specification of space required in individual cells
	Material Handling	Specification of material handling requirements
	Factory Storage	Specification of factory storage requirements
	Support Services	Specification of support services required
	Factory Facilities	Specification of factory facilities required
	Supply Chain Structure	Identification of suppliers and customers
	Supply Chain Modeling	Visualization of logistics network
	Facility Location Planning	Location of manufacturing and distributing facilities
Human and Organization	Organization Structure	Specification of type of structure of the MS organization
	Organization Culture	Specification of culture required for the MS organization
	Organization State	Specification of operating conditions for the MS
	Labor Policy	Specification of labor policies to be adopted
	Quality Policy	Specification of quality policies to be adopted
Information and Control	Integration	Specification of degree and extent of integration of identified entities
	Autonomy	Specification of degree and extent of autonomy of entities
	Automation	Specification of degree and extent of automation of identified entities
	Planning and Control	Specification of planning and control functions
	System Architectures	Specification and modeling of information and decisional architecture
	Data Flows	Identification and modeling of major information flows

- *Information and control.* The purpose of this task frame is to specify the informational entities of the MS system at a conceptual level of detail. Hence, it is concerned with the specification and modeling of the MS management system, the degree of autonomy and independence for decision making and the flow of data within the systems.

- *Consolidation—system feasibility study.* The final section of the conceptual design phase in the MSD process is the consolidation stage. The purpose of this is similar to the previous consolidation stage: to bring together all the design requirements and ideas captured, created, and generated within the three task frames. The result should be a conceptual design model that is static in nature. The MS system should be defined in a structural and functional sense, and the requirements for each MS unit should be specified. Finally, an MS systems feasibility study report is generated for the approval of the MSD steering committee. This report should ensure the consistency of the conceptual system design specification with the overall business and MS strategy policies and the system requirements. Based on the concepts developed in the conceptual design stage, the feasibility study report should identify the structural, functional, financial and managerial feasibility of the conceptual system design and the MS sub-architectures.

Table 1.3 Processing MSD tasks

	MSD Task	Description
Processes	Detailed Layouts	Specification and design of layouts of individual factory domains
	Detailed Cell Layouts	Specification and design of layouts of individual manufacturing cells
	Workstation Layouts	Specification and design of layouts of individual workstations
	Equipment Selection	Specification and selection of individual items of equipment
Facilities	Human Services	Specification and design of services required for employees
	Material Services	Specification and design of services required for physical materials
	Machine Services	Specification of services required for machines and equipment
	Buildings	Specification and design of the building
	Health and Safety	Specification of environmental health and safety issues
Supports	Maintenance	Specification of maintenance policies and functions
	Tooling	Specification of tooling policies, functions and location
	Supplies	Specification of supplies policies, functions and location
	Administration	Specification of cell level administration policies, functions, roles
	Setup Management	Specification of setup management policies, functions, and roles
	Process Inspection	Specification of inspection policies, functions, and roles
Plan	Production Planning	Specification and detailed design of production planning functions
	Scheduling	Specification and detailed design of scheduling functions
	Software Definitions	Design/selection of software for production planning and scheduling
	Equipment Selection	Specification of soft/hardware for production planning
	Batch Sizes	Specification of optimum batch sizes and range of batch sizes
	Volume Mixes	Specification of optimum volume mixes and range of volume mixes
	Shift Patterns	Specification and design of shift patterns
Control	Control Systems	Specification and detailed design of MS control systems
	Data Collection	Specification and design of data collection methods and techniques
	Materials Management	Selection/design of materials management techniques
	Software Definition	Specification and design/selection of software for control systems
	Equipment Selection	Specification and selection of types of control systems equipment
Human	Job Requirements	Specification and analysis of job requirements
	Job Design	Specification and design of jobs, roles and responsibilities
	Training	Analysis of training requirements and specification of training program
	Quality	Specification and design of quality systems, roles and responsibilities
Organization	Structure	Specification of organizational structure
	Working Conditions	Specification and design of working environment
	Safety	Specification and verification of safety issues
	Motivation	Specification and design of workforce motivation methods
	Reward Systems	Specification and design of reward systems
Warehouse	Buffer Sizes	Specification of buffer sizes
	Storage Locations	Specification of location of WIP buffers and storage areas
	Storage Systems	Specification and selection of types of storage systems
	Handling Paths	Specification of handling paths
	Handling Units	Specification and selection of types of handling units
	Warehousing	Specification of warehouse design and management
	Transportation	Specification of inbound, intermediate and outbound transport

Detailed design

Within the detailed design stage, a number of alternative MSD options are generated and assessed based upon the conceptual design developed previously. The detailed design stage represents a more in-depth investigation of the three sub-architectures. It is based upon the development of a series of sub-systems that directly contribute towards the operations of the MS system. It aims to identify a number of approaches that may fulfill the system's requirements. As such, it needs to take into account the existing system's structure and functionality, as well as any constraints imposed by the existing system. It comprises the main MSD tasks as given in Table 1.3:

- *Processes.* The purpose of this task frame is to specify the physical aspects of the MS processes in detail. Hence, it is concerned with the physical processes and the selection and positioning of equipment.
- *Facilities.* This specifies the individual service requirements that the factory needs to provide.
- *Support.* This specifies the location and operating policies for activities that support the MS operations within the individual cells.
- *Planning.* This specifies the planning and scheduling functions and operating policies of the MS system.
- *Control.* This specifies the control functions and operating policies of the MS system.
- *Human.* This specifies the design requirements specific to human issues.
- *Organization.* The purpose of this task frame is to specify the design requirements specific to organizational issues.
- *Warehouse and transport.* This specifies the transportation and materials handling equipment required, as well as the storage equipment and locations.

1.4.4 MS System Implementation

This specifies the functionality and procedures of the MSM phase of implementation, dealing with two closely related areas: system implementation and system change management. In general, a coherent set of detailed plans and instructions should be prepared to effectively manage the necessary future changes. An implementation plan should, for instance, include items such as an outline of the requirement of change, a description of method of change, a specification of the tasks and resources required, and a time plan for the implementation project. The aim is to help achieve the project goal smoothly, in the shortest possible time, and at the minimum cost. Once started, the progress of implementation tasks will need to be continuously monitored. If necessary, feedback actions should be taken to adjust the actions being taken. Eight main components can be identified as essential for accelerating change and maximizing its chance of success (Figure 1.15).

These components of change management provide the basis for the structure of the MS implementation phase of the MSM framework. The aim is to link the new system design, developed during the MSD phase, into transition plans and implementation programs which will lay a foundation for a successful implementation of the new system. Again, the three main aspects that are incorporated in the implementation phase are processes, information technology (IT), and organization and human resources. This phase will take the outputs of the MSD phase as inputs. It begins with the stage of *preparation for change* to provide a basis for the development of transition plans, which include scheduling, budgeting, and resource requirements. These plans are the basis to bring the new manufacturing/logistic system design into reality.

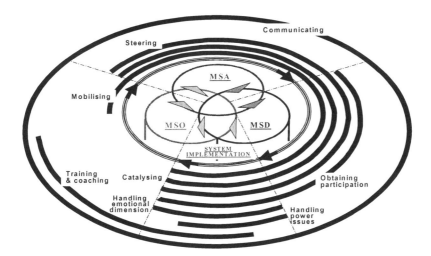

Figure 1.15 Steps of change management within the MSM context

A three-stage procedure has been developed for MS implementation (Figure 1.16):

- *Stage 1—Preparation for change.* This aims to make sure that the organization is ready for the changes required by the MSD initiatives, so that the MSD team and the system user have a common understanding of all the definitions used in the design. Everyone concerned should be motivated by the strategic vision.
- *Stage 2—Transition plan development.* The second step in the implementation phase is to develop one or more transition plans. A transition plan includes project time plans, resource allocation plans, budgets, performance measures and contingency plans. The alternative transition plans are compared and evaluated, and the most favorable for a successful implementation of the new system is selected. Initially, project scheduling is done to allow planning of the activities. This is followed by resource management and project budgeting. After an iterative process of transition plan refinement, process performance measures are then selected. Finally, the complete set of the previously specified independent projects is integrated into a master transition plan.
- *Stage 3—Implementation.* This step consists of planning and performing the actual implementation of the new system. This includes monitoring and controlling the progress of the implementation, and evaluating the success of the project.

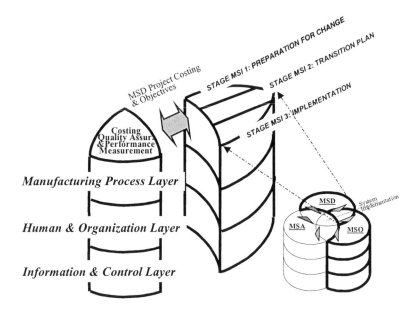

Figure 1.16 Stages in MS implementation

1.4.5 MS Operation And System Status Monitoring

MS system operations are an established functional area of manufacturing with well-developed theories and tools. The current development of Enterprise Resource Planning (ERP), which inherits its nature from its forerunner, Manufacturing Resource Planning (MRP II), is a typical example of the kind of IT systems used to provide an integrated information system for the planning and control functions required. With the move toward IT integration through client/server and Internet, ERP is being pushed from a conceptual to a practical arena. A number of unsuccessful cases reported in the literature, however, show that purely technical-oriented ERP implementation is one of the main reasons for failure. There seems to be a lack of a structured, strategically driven approach to assist companies mapping function-oriented software onto a business-oriented system. It is evident that different industrial companies have different focuses on their business and manufacturing function. Current ERP systems also have different merits and weaknesses when related to different industrial requirements. The proposed MSM framework provides a sound basis for a strategically driven analysis of MS information system requirements, giving a strategic direction for information system evaluation, implementation, and administration.

In particular, the system performance monitor is needed to complete the MSA-MSD-MSO-MSA cycle. This area is particularly important for the framework's real-life adaptation and operation. This is because it is responsible for the continuous monitoring and reporting of the current system performance against the pre-established strategic goals. In accordance with the pre-conditions for efficient systems operation, MSM performance measurement is generally needed to:

- provide the MSM system with a method to assess its current competitive position with respect to its current strategic direction, its competitors, and the demands of the market, and
- monitor the system's progress towards its strategic objectives and identify avenues for continuous improvement.

In addition, external influences should also provide a stimulus to the initiation of the MSA-MSD-MSO cycle. Being an open system, the company cannot otherwise be certain that objectives established for future improvement will be adequate to lead to superior competitive performance. This can only be achieved by evaluating and quantifying the current state of the company, by highlighting where improvements have been made, and by defining areas which need improvement. By using performance measures that are supportive to a company's strategy, the feedback from the process provides the company with the information needed for ongoing improvement. It allows for monitoring of the critical success areas and points out which corrective actions to take should a drift occur. Therefore, this MSM performance monitoring module aims to monitor and initiate the right action whenever and wherever necessary in the MS processes.

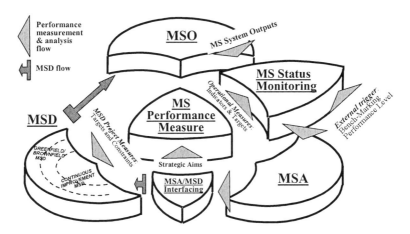

Figure 1.17 Overall structure of system status monitoring

Various approaches have been suggested for performance measurement dealing with performance issues at each of the three MS layers. However, within the context of MSM, an extended scheme of evaluation is required so that the key requirements can be addressed (Figure 1.17). The performance-monitoring module is closely related to the MSA process, with a certain degree of overlapping between the two. In order to ensure that an MS system achieves a strategically competitive position, and that different parts of the organization are pulling their weight in a combined effort to maintain this position, some form of coherent performance monitoring is essential. This monitoring must be applied to individual units, as well as to the whole organization. The ultimate aim of performance measurement is to motivate behavior leading to continuous system improvement. When integrated within the MSM framework, the monitoring module has the following features:

- Within the MSM framework, it provides a mechanism of closed-loop for both the monitoring and the continuous improvement of the system.
- It is completely integrated with the MSA domain. Strategic concerns are disaggregated into operational level measurements in a top-down manner. Then, the actual operational level measurements are aggregated back, following a bottom-up approach, to reflect the system's performance against its current strategic goal.
- It is dynamic in nature and, together with the system audit approach adopted by the MSA module, allows the systematic revision of critical areas, performance measures, historical data, decisions, and outcomes.
- Both the present performance requirement (based on an internal gap analysis) and predicted future requirement (based on an external gap analysis) can be taken into consideration.
- Both global optimization (through an overall MSA/MSD process) and local optimization (through continuous-improvement MSD action plans) can be supported.

1.4.6 Task-Centered MSM Workbook

According to the structure and processes of the MSM framework presented in the previous sections, a complete workbook has been developed. This workbook, which is presented in the subsequent chapters, provides step-by-step guidance through the MSA-MSD-MSO cycle.

The workbook is structured in a task-centered way. *Task-centered* is the concept of providing all the information relating to a particular task at the point where the task is to be executed, allowing the user to navigate through the processes as required, and to access the relevant information in a focused way. Necessary elements, such as task description, instructions, processes, drawings, tools, and data are all assembled and integrated into a single working page, and presented as a single entity known as a *task document* (Figure 1.18). For each task document, four additional types of work sheets can be provided to aid in the execution of the tasks:

- Questions and data collection sheets, to assist the development and the capture of the individual MSM decisions.
- Cause-effect linking tables to assist the association of MS strategy concerns to the necessary system design actions, as described previously.
- Tool sheets to specify relevant tools to be used (e.g., graphical, analytical, computer-based), and the related inputs and outputs.
- Checklists to help identify the relevant issues to be considered during the analysis/design processes. An example of this within the MSA/MSD interface module is a 'quick hit' table. This provides an indication of some of the typical

problems prevalent in each of the MS policy areas and their effects on the competitiveness of the MS systems with respect to six key competitive criteria, and vice versa. Other examples include checklists for change management issues and, where appropriate, for some of the key MSD tasks.

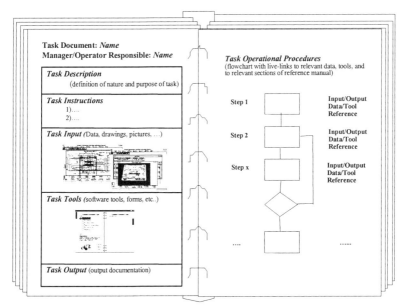

Figure 1.18 The structure of an MSM task document

1.5 CASES OF INDUSTRIAL APPLICATION

The following cases illustrate how the MSM framework can be applied in industrial settings. The MSD projects involved include both greenfield and continuous-improvement types.

1.5.1 Case A: From Strategic Initiatives to System design Action Plans

The first case study involved a major UK high-tech manufacturing organization. From a strategy analysis performed on a potentially profitable element of the business, a series of action plans and associated MSD projects were identified as a means of contributing towards the improvement of the manufacturing function. These were divided into three categories: organizational issues, such as changes required in company culture; quality issues, such as the need for proper documentation to increase traceability and control; and other MSD issues, such as those related to relocation of product based manufacturing cells within the factory.

In particular, this case study highlighted some further issues with respect to the implementation and application of the MSA/MSD interface. The procedures contained within the interface model were found to be useful within the company's strategy-planning group. Having prioritized the decision areas to be addressed, the interface model provided an additional verification of the consistency and completeness of the strategy by suggesting associated decisions that would otherwise have been overlooked.

The capturing of the strategy and the ability to retrieve the decisions and the rationale behind those decisions was one of the important benefits identified by the company's strategy group. It was felt that, compared with the existing approaches that leave the companies almost entirely on their own at this stage to identify feasible options, the MSA/MSD interface equipped the users with a structured guide to enable them to make more informed decisions. The results were seen as being an improvement on the company derived project structure that was considered to be of too high a level of abstraction for effective application and implementation.

1.5.2 Case B: Design of a Greenfield MS System in the Automotive Industry

The merging of automotive manufacturers highlights another application area where a framework such as the MSM is needed. Maximizing the benefits of such mergers requires the effective convergence of the organizations' processes, which is a complex undertaking that requires a structured approach. An approach known as business process development (BPD) was used in the design of a major European car manufacturer's new engine factory, illustrating how the MSM framework can be applied to deal with a range of issues related to the analysis, design and implementation of a new manufacturing system. It also shows how being an integral part of the MSM framework enables the system to be continually reengineered in accordance with environmental changes.

Strategic background

The increased competitive pressure within a globalized automotive industry has led to mergers and acquisitions by many manufacturers. The benefits expected from these are:

- Shared research and development costs/competence.
- Economies of scale in material costs.
- Bargaining power against major suppliers.
- Increased manufacturing flexibility.
- Reduced dependability on local economic cycles.
- Expansion of brand/market sector coverage.

In the case of the example company's new European engine factory, a number of strategic drivers existed. These derived from the group's acquisition of another organization in the mid-1990s. To achieve the business objectives of this acquisition, the product strategies of both organizations had to be aligned. For instance, it was decided to pursue a common engine strategy, where families of "new generation" engines would be designed for the complete range of vehicles.

To deploy this product strategy, the manufacturing strategy of a global production network had to be implemented. A decision was made to build a greenfield engine factory that would manufacture a range of four cylinder petrol engines, producing an annual volume of up to 500,000 engines with a workforce of about 1,500. Volume production commences in early 2001. For this factory to fit into the group's production network, many of its engineering, logistical and business processes had to interface to processes within the network. Hence, they were required to share functional commonality with those in other engine factories. Following this strategic guidance, it was decided to design the new factory according to a business systems model based on a model plant. The following issues were raised:

- How does one analyze a complete system, including the actual processes, organizational structures, IT systems and the underlying qualities of the processes (i.e. the "soft" factors)?
- How does one structure the redesign of the business system to ensure the completeness of the total system and the fit of MS processes within the system?
- How does one ensure the timely implementation of the system, in line with the introduction of a new product and the build up of a new organization?

The questions became even more important when considering the size and complexity of the business system of a highly automated engine manufacturing facility.

Figure 1.19 Product of the new engine plant

Conceptual MS system architecture

To cover all functional areas, the system needed a hierarchy of processes. These processes ranged from the design, manufacture, assembly, and delivery of the product, to support processes such as quality management, finance and controlling, personnel management, facilities management, and so on. It became apparent that a structured approach was needed to enable the project team to analyze and evaluate the existing system model. The design and implementation of the new, improved processes would need to proceed in a timely manner. In close relation to the overall MSM framework, the BPD process adopted by the company had four major steps:

- Business process analysis—to analyze or learn the model processes.
- Business process evaluation—to evaluate their strategic fit and their strengths and weaknesses.
- Business process design—to design complete business processes and a complete business system.
- Business process implementation—to implement the processes and train the relevant people in a timely manner.

To support these steps, two models were used as the backbone of the BPD process: the MS processes and the MS systems. In accordance with the conceptual MSD framework, the MS system model enabled the structuring of the overall business system, as shown in Figure 1.20. It provided guiding principles in terms of internal customer-supplier relationships and a visual design tool. Such a model enabled the top-down design of the business system as well as the capturing and structuring of bottom-up process design activities.

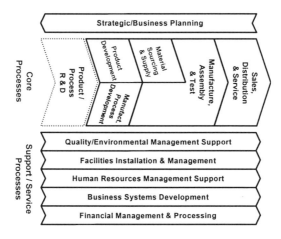

Figure 1.20 MS system model

At the detailed MSD levels, the process model defined all the elements of an MS process, as illustrated in Figure 1.21. The model begins with the internal or external customer of the process, who defined the critical success factors (CSFs) of the process, and the performance measures derived from the CSFs. Therefore, this model closely followed the generic conceptual structure of an MS system architecture as presented in Section 1.3.3. That is, it specified the process or set of activities to achieve the CSFs; the organization structure to operate these activities; the people and their competencies within this structure; the IT systems to support information flow, processing and storage within the process; the facilities and equipment; and the infrastructure requirements of the process.

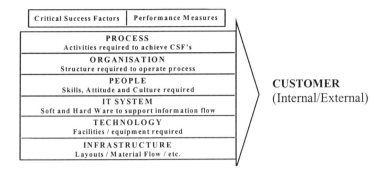

Figure 1.21 MS process model

When analyzing an MS process, all its elements must be analyzed and understood as a whole (Figure 1.21). In addition, all of the elements had to be included and aligned to one other. The process would be unlikely to achieve the desired outcomes if these were not satisfied. Therefore, the framework provided a mechanism to categorize the interdependent components of an MS process. It was used to structure process analysis, evaluation, design, and implementation. In this particular case, the hierarchy of processes contained eleven high-level processes, which could be broken down into seventy distinguishable MS processes. These processes could be broken down further into about three hundred sub-processes. The BPD process started with the formation of a BPD team for each high-level MS process of the plant. The team was led by a process owner and contains members from both customer functions of the process, and inputting/executing functions of the process. This team was responsible for the delivery and ongoing management of an improved MS process throughout its life. The analysis of the model business system and its MS processes had three essential considerations:

- Structuring of the analysis or learning process to ensure total coverage while avoiding duplication.
- Comprehension of the complex system of processes and the complexity of processes themselves.
- Understanding of the key question: "what makes it work?"

The first challenge was met by using a quality management system (QMS) of the model plant as the analysis structure (Figure 1.22). The QMS is a description of all processes—about three hundred hierarchically structured procedures. The business process model was used to aid the comprehension of a process and to structure the actual analysis of a process. The last challenge required "living" the process, meaning to work in the process and its organization for a significant period of time.

	BMW Steyr process (QSV)	BMW Steyr Dept	BMW Steyr Name	NG4-F Dept	NG4-F Name	Name	CSF's	Proc. Flow	Org. Struct	Skills & Att	IT Syst.	Tech.
									Business Process Components			Project Status
QMH	QMH											
QMV 05.01 ZM	Producing QSVs, QSAs, and	ZM-Q-10	Hollnbuchner	2339	Quality	R. Desai	36235	x	xx	y	x	x	xx	N/A
QSV 04.01 ZM	Targets	ZM-Q-10	Hollnbuchner	2339	Quality	R. Desai	36235	x	xx	y	x	x	xx	N/A
QSV 05.02 ZM	Quality audits	ZM-Q-10	Hollnbuchner	2339	Quality	J. Cumiskey	35689	x	x	y	y	x	x	N/A
		ZM-Q-10	Hollnbuchner	2339	Quality	R. Desai	36235	x	x	x	y			
QSV 05.04 ZM	Product audit	ZM-Q-20	Furthbauer	2400	Quality	J. Marsden	33718	x	x	x				
QSA 05.04.01 ZM-Q-20	Dynamic audit (engine analysis)	ZM-Q-20	Pichler	3139	Quality	J. Marsden	33718	x	x					
QSA 05.04.02 ZM-Q-20	Dynamic audit (performance test)	ZM-Q-20	Tursch	3139	Quality	J. Marsen	33718	x	xx	x				
QSA 05.04.03 ZM-Q-20	Static audit (engine audit)	ZM-Q-20	Eidenberger	3307	Quality	J. Marsden	33718	x	xx	x				
QSV 05.05 ZM	Roles & Responsibilities regarding Quality issues	ZM-Q-10	Hollnbuchner	2339	Quality	D. Foster			y		y			
QSV 06.01 ZM	Economics and cost of poor quality			2338	Quality	D. F.			y					
QSV 09.02 ZM	Quality related activities with suppliers			2204	Quality	P. Collins			xx	xx	y	xx	y	x	Y
QSA 09.02.01 ZM-T-40	BOF/ BOR performance test		er	2526	Quality	P. Collins			y	x	y	x		N/A	
QSA 09.02.03 ZM-T-40	AQUA			3231	Quality	P. Collins			y			x			
QSA 09.02.04 ZM-T-40	Procedure for production transfers							x	y	y				Y

Figure 1.22 Sample matrix of MS process analysis matrix

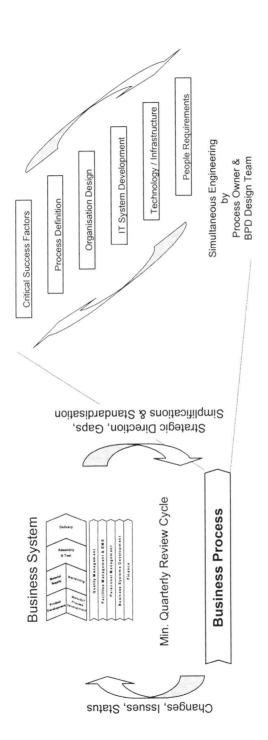

Figure 1.23 Two levels of design—systems and processes

The QMS of the model plant combined with the process model lead to the MS process analysis matrix. This matrix aided the project management of the analysis process at the actual process level and the overall systems level by visualizing what had taken place, and by highlighting areas needing further analysis, as illustrated in Figure 1.23.

Overall MSA/MSD task and reference structure

As indicated in Figure 1.6, the MSM framework essentially supports a structured mechanism for both the execution and the communication of system designs. Therefore, in addition to analyzing the processes of the model plant according to the generic MS system architecture, the BPD teams must also evaluate these by carrying out three activities:

- Strategic fit evaluation—model plant process performance vs. strategic targets of the new plant.
- SWOT analysis—identification of strengths and weaknesses of the model process, as well as opportunities and threats of re-implementation in the new plant.
- Specific requirements—the new environment may differ from the model plant requiring a process change.

The above would produce the input for the development of the critical success factors, the first action of the actual process design. Following the evaluation, design was carried out at two levels (Figure 1.23): the MS system level and the MS process level. The MS system level design ensured the completeness of the overall system, the fit of processes and the strategic direction of the system. It also identified opportunities for standardization and simplification. At this level, the management team reviewed the process design work on at least a quarterly basis, using the business system model as a structuring and graphical tool. At the MS process level, the BPD teams designed individual processes or process groups following the MS process model. The starting point was the definition of the critical success factors of the process. This was followed by the actual specification of all the elements, as outlined before, to ensure completeness of the design. The design process itself was of a simultaneous nature, ensuring the overall fit of the MS process and the fit to its process/systems interfaces. One of the outputs of the design work was the quality management system of the new plant. The procedures and instructions were produced in parallel to the design work, thus aiding the design process by making it more objective.

Project management and system implementation

The implementation of an MS process covers all of its components, as described in the *MS Implementation* module. Performance measures must be implemented, the process has to be communicated, and people trained. In addition, the organizational structure must be established (including the relevant management control structures), IT systems have to be implemented, and facilities have to be installed and commissioned. Hence, the timely design and implementation of the system require project management, in addition to the systems engineering elements. The situation in the case study was that approximately seventy MS processes owned by about fifty process owners had to be designed. The number of people involved in the design was estimated as between five and twenty people per process, with many of these being involved in more than one process design. Therefore, the number of people involved in the design of the processes reached up to two hundred. To manage and control these tasks, an effective organization and management control structure was required. The key role in this organization was the process owners, who were responsible for making all the activities take place, and for achieving the customer requirements of the processes.

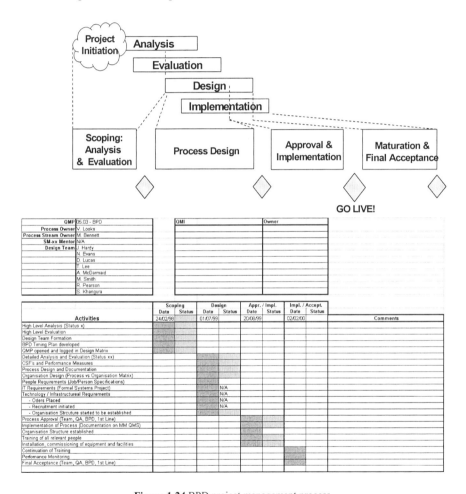

Figure 1.24 BPD project management process

As shown in Figure 1.24, the backbone of implementation here was based upon effective communication and extensive training of all relevant people in the process. Professional training developers were involved to facilitate the process design teams in the development of training programs and their execution.

Maturation of the implemented processes was also an important factor for success. The faster the processes became embedded in the conscience of the organization, the faster the organization would reach its performance targets.

To achieve this maturation quickly, a number of things must happen. Firstly, the process design work had to be a shared team effort by representatives of all functions, which had a stake in the process. This increased the likelihood of functional acceptance of the designed process as well as the fit to other processes owned by the involved functions. The implementation was then mainly a matter of rolling it out to the wider user population, and not a matter of lobbying customers of, and contributors to, the process. Secondly, initial and ongoing training had to support the rolling-out of the processes. This included extensive coaching and facilitation by the process owner and key personnel in the process, especially in the early phases of the implementation. It had to be recognized that training alone will not create competence in the operation of a process. A learning curve had to be mastered. The aim was to reduce the duration of the learning through a logical combination of training and coaching.

Since many of the processes designed were interdependent, the components of an MS process could link to many other activities within the project. Hence, the timing of the design of one process had to be aligned to other relevant project activities. This required that the decisions made during the design phase be continually reviewed to ensure the coherence of project activities.

The major tool developed for this task was known as the BPD design checklist. The execution of the BPD process for each MS process was controlled by a single checklist that captured all project management information: process ownership, the design team, the scope of the process, and all of the activities to be carried out. The activities of the BPD process were grouped into four distinct phases, with reviews held at the end of each phase. The review of the 'approval and implementation' represents the 'go-live' point of the MS process (Figure 1.24).

Figure 1.25 Engine assembly line

Case observations

The case of the design and implementation of a greenfield engine plant clearly demonstrates that the design and management of a manufacturing system is a complex domain. Without a logical framework and its associated tools, such as the MSM, a coherent and strategically oriented system could not be designed and implemented in time. In particular:

• The logical structure of the MSM framework helps to set systems thinking into the context of manufacturing systems management, by helping an organization identify the key functional areas, outline the contents and relationships within them, and then logically integrate them into a closed-loop to provide the basis for the development of a set of consistent parameters and procedures.

- Following the above, the design of processes within manufacturing requires a simultaneous engineering approach where experts from the various elements of a functional area work in parallel to define the optimal total solution for the MS process within the overall system.

- Although the idea of the internal customer-supplier relationship within an organization has existed for a significant time, there were still functional "kingdoms" which did not like to be told by others (the internal customer) what to do. The structures put in place within the BPD process, however, forced these functions to involve their customers, creating the willingness to discuss the CSFs of the MS process with other functions.

- Process ownership was another area where the approach of the BPD process brought significant learning. Historically, there was no real process ownership within the organization, in the sense of making the process happen. The important thing is that a process owner should not only be the person who wrote the procedure describing the process, but he or she must also be responsible for all the relevant activities as specified by the generic MSM framework, and make things happen. This turns process owners into quite powerful members of the organization. It also shifts some power from functional managers or senior management to process owners, which are usually junior management. In other words, the power shifts from an almost purely managerial level to a "doing" level in the organization. This leads to an empowerment of a level in the organization, which in the past was mainly the executor of senior management's decision.

- The design of an MS function has many dependencies to other activities, as an MS process will normally be linked by all of the three layers as shown in Figure 1.6. Hence, the timing of all these activities has to be aligned to avoid decision-making that would create limitations for other dependent decisions.

1.5.3 Case C: Development of a Strategically-Driven MIS

The implementation of a manufacturing information system (MIS) within a manufacturing organization often forms part of the strategic approach to satisfying manufacturing requirements. This case addresses the link between manufacturing strategic issues and the requirements of MIS structure and implementation. Following MSM's structure of evaluation, a set of MSD tasks was specified within the information and control task frame, dealing with initial identification of objectives, available systems analysis, "develop or buy" decisions, structure design, and implementation. The approach has been applied successfully to the case of a typical modern precision engineering company. The company heavily utilizes computer numerically controlled (CNC) facilities and specializes in the making of aerospace and telecommunication components. It offers services from prototypes only, through production batches. Through an analysis of the company's manufacturing strategic requirements, the proposed procedures revealed a number of MIS-related issues and features that helped to ensure a competitive edge.

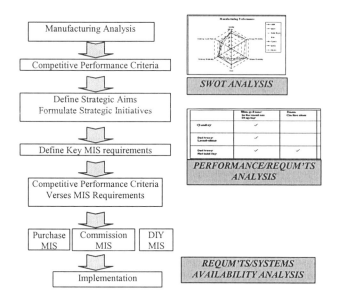

Figure 1.26 Overall process

Manufacturing strategy and MIS

The development of enterprise resource planning (ERP) inherits its nature from its forerunner, manufacturing resource planning (MRP). ERP is a typical example of the kind of IT systems used to provide an integrated information system for the planning and control functions required. However, it has been observed from a number of unsuccessful cases reported in the literature, that the purely

Figure 1.27 Machined parts of the example company

technical orientation of ERP is one of the main reasons for its failure. There seems to be a lack of a structured, strategically driven approach to assist companies mapping function-oriented software onto a business-oriented system. It is evident that different industrial companies have different focuses on their business/manufacturing function. Current ERP systems have different merits and weaknesses, when related to different industrial requirements. The proposed MSM framework provides a sound basis for a strategically driven analysis of manufacturing information system requirements, giving a strategic direction for information system evaluation, implementation, and administration. At the information and control level, in particular, the normal process of manufacturing strategy analysis is extended by adding a set of generic procedures. These procedures help companies identify key MIS and system requirements based on the initiatives derived from strategic analysis. This strategically driven analysis approach aims to identify the key MIS requirements required in order to satisfy any designated competitive performance criteria.

As summarized in Figure 1.26, each of the whole processes can be divided into three sections: the definition of manufacturing strategy aims and initiatives (starting with the MSA process carried out against the competitive performance criteria, with the polar plots drawn for each of the customers/products, leading onto the definition of the strategic aims through a SWOT analysis), the identification of key MIS requirements (cross reference via tabulation drawn of competitive performance criteria versus key MIS requirements), and the decision on the choice of MIS design, structure and implementation (either through the purchase of an off-the-shelf system, a customized system or by in-house development).

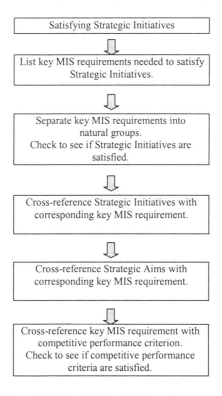

Figure 1.28 Identification of MIS requirements

Each stage of the generic procedures will be identified and presented in simple terms, allowing the user to gradually progress through the stages. For instance, one of these requires tabulation of the key MIS requirements and the corresponding strategic aims. This correlation can serve as a reminder of which of the initially defined strategic aims has been instrumental in establishing the particular key MIS requirements. To help this process, the user may employ a set of generic correlations between the competitive performance criteria and key MIS requirements, with cross-checking, as illustrated in the flowchart of Figure 1.28.

Market analysis and manufacturing strategic initiatives

The subcontracting marketplace has a reputation for being tough and competitive. Although the reasons for subcontracting have not changed, many organizations now regard their subcontractors as an important extension to their own facilities, taking the necessary steps to make them feel part of their team. This has resulted in organizations reducing their supplier base by selecting the companies that they feel can offer the best service. With this reduction of suppliers within companies' supplier bases, comes even more fierce competition. This competition comes not only within the same supplier chains, but also globally, with subcontractors wishing to be included within the supplier chain of an organization.

Table 1.4 Summary of gap analysis result
(W: Order winning, which significantly contributes to winning business; P: Potentially order winning; Q: Order qualifying, those aspects of competitiveness where performance has to be above a certain level even to be considered by the customer)

Criterion		Co. A	Co. B	Co. C	Co. D	Co. E	Co. F	Co. G
Quality	Gap	-10	10	-10	-10	10	10	10
	Qualifier	Q	Q	Q	Q	Q	Q	Q
Lead-time	Gap	0	-20	-10	-10	-40	-30	-30
	Qualifier	W	W	W	W	W	W	W
Lead-time	Gap	-30	-30	-60	-50	-20	-20	-20
Reliability	Qualifier	W	W	W	W	W	W	W
Design	Gap	10	70	80	70	-10	10	30
Flexibility	Qualifier	P	P	Q	Q	W	Q	Q
Volume	Gap	10	0	20	30	-10	-10	10
Flexibility	Qualifier	W	W	Q	Q	Q	Q	Q
Cost/Price	Gap	30	40	0	0	50	-10	-10
	Qualifier	P	P	P	P	P	P	P

In order to increase its competitiveness, a customer survey was carried out by the company to determine its customers' requirements, and to identify how orders are won against competitors. Table 1.4 summarizes the performance gap for each of the company's key customers. The possible range was –100 to +100, with a positive number implying that manufacturing performance criteria has been exceeded, and a negative number implying performance needs to be improved. In particular, it was revealed that for both Delivery reliability and Delivery lead-times, almost all the results showed negative gap values. In this particular case, Delivery lead-times could be further divided into Delivery lead-times for production, and

Delivery lead-times for the manufacture of prototypes. Both these numbers would need to be reduced in order to remain competitive.

Table 1.5 Sample strategic aims/initiatives table

Competitive Criterion	Strategic Aims	Strategic	Initiatives
Delivery reliability	Improve Delivery reliability and predictability.	Consider finite capacity of personnel. Consider finite capacity of machine tools.	Give operators explicit instruction. Monitor job progress constantly.
	Create stability.	Eliminate unknowns through improved planning.	Implement preventative and planned maintenance.
	Provide information to minimize time waste.	Implement shop floor MIS that provides all necessary operator information.	Provide information on tooling and fixture setup with written and visual aids. Provide integrated information package.
	Establish accurate standard times.	Implement MIS to monitor setup and cycle times and to re-establish standard times as necessary. Monitor delivery performance.	Improve time estimates by referring to historical manufacturing information and collected data.
Delivery Lead-times (Production)	Eliminate time waste.	Monitor machine tool performance. Collect time and attendance data. Provide correct information.	Provide full documentation of proven, reusable manufacturing methods. *(Not "reinventing the wheel".)*
	Reduce production lead-times to less than that of competitors.	Establish lead-times with customer. Use customer CAD files for drawing modifications to aid re-programming speed and accuracy.	Reduce lead-times by accurate capacity planning. Reduce lead-times by concurrent manufacturing.
	Encourage customers to provide design change information direct from CAD.	Demonstrate speed and cost-saving advantages.	Demonstrate information integrity and reduced prove-out time.
Delivery Lead-times (Prototype)	Eliminate time wasting.	Monitor machine tool performance. Collect time and attendance data.	Provide correct information. Create tooling visual display.
	Reduce prototyping lead-times to less than that of competitors.	Use customer CAD files to aid programming speed and accuracy.	Recall historical data of similar parts or features.
	Encourage customer–supplier information exchange.	Demonstrate benefits of early design information.	Value engineering (to reduce both time and cost).

However, it could be argued that it is more important to reduce lead-times of prototype components, since these are nearly always needed in a hurry. Furthermore, the supplier selected to build the prototype is frequently the supplier that ends up manufacturing the production run. It is therefore important to understand and to find ways of improving delivery performance, especially for prototyping operations. For instance, it is generally much more difficult to prepare a prototype component than to prepare a component that has previously been manufactured. Time benefits may be gained by using computer-aided design (CAD) file information directly from the manufacturer's computer-aided manufacturing (CAM) system, assuming the customer allows this transfer of data (which is more likely if the customer benefits from the reduction in lead-times and possibly in cost). By making such a gap analysis for each of the criteria, the company identified its future strategic aims/initiatives under each of the headings. A sample of these is shown in Table 1.5.

Key MIS requirements

To specify the MIS requirements, which may affect the defined strategic initiatives, it is essential that there be a clear understanding of exactly what the strategic initiatives are. This ensures that valid judgment is then made as to whether the strategic initiatives will be achieved by the proposed solution. In considering the MIS requirements for satisfying strategic initiatives, the appropriate MIS features for each functional group should be taken into account. While the list of appropriate features for each of the functional groups (Figure 1.29) is not extensive, it does serve as a foundation on which to build:

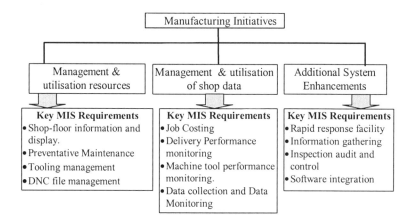

Figure 1.29 Identification of key MIS requirements

- *MIS features for the utilization of plant and resources*. The four basic MIS features that were selected for improved utilization of plant and resources were shop floor information display, machine tool preventative maintenance, tooling management and computer numerical control (CNC) file management. These features were selected as they covered most aspects of plant utilization. However, it was accepted that MIS features or requirements could be supplemented indefinitely until any given strategic initiative was satisfied. Another reason for selecting these basic MIS requirements was that they were broad in definition and covered a wide range of material within the topic. For instance, CNC file management could include programming and editing aids for the production of part programs as well as the ability to transfer part programs between machine tools and the programming office.
- *MIS features for the utilization of collected data*. The four basic MIS features that were selected for improved utilization of collected data were time and attendance monitoring, delivery performance monitoring, machine tool performance monitoring and job costing. Again, these features were selected because they covered most aspects of data collection. It was also accepted that MIS features or requirements could be supplemented indefinitely until the strategic initiative was satisfied.
- *MIS features for the additional system requirements*. The four basic MIS features that were selected for additional system requirements were rapid response facility, information gathering, software integration and inspection audit and control. These MIS items were used to illustrate the diversity of available features. The selection of additional system requirements was seen as a spillover from the utilization of plant and resources and the utilization of collected data. In this case, an MIS with a rapid response facility had the features that were required to assist in providing a manufacturing rapid response service along with normal production controlling systems. Similarly, an MIS that provided information gathering could be explained as having the mechanism to manage the accumulation of data from information gained throughout the production life cycle for any given component. Although these MIS requirements were somewhat diverse and non-intuitive, they served to illustrate the purpose of this particular functional group.

It was next necessary to check each of the initiatives in turn to see if the basic MIS features were able, in principle, to satisfy them. By definition, this would have the desired effect on the relevant competitive performance criteria. In the case of the company, this helped to establish twelve key MIS requirements (Table 1.6).

These acted as a quick reference to identify the strategic initiatives that instigated the particular key MIS requirement. This, in turn, allowed management to evaluate available MIS systems based on their strategic requirements, as illustrated in Table 1.7 (N.B.: this table is for demonstration purposes only—it has no general implication regarding the features of any specific system). Through this analysis, the company identified two major areas where key MIS requirements had not been met by any of the systems available (rapid response facility and job costing) and hence, the corresponding strategic initiatives that could not be directly supported. Owing to the implications of these inadequacies, the company decided to make a purpose-built system that more closely supported the requirements.

Table 1.6 Key MIS requirements and corresponding strategic aims

Requirements	Strategic Aims
Shop floor Information and Display	Promote information availability throughout the manufacturing process. Improve small batch handling through reduction of programming prove-out time. Improve small batch handling through setup time reduction. Encourage customers to provide any design changes direct from CAD. Eliminate time wasting. Improve Delivery reliability and predictability. Provide information to minimize time waste. Improve standards above those of competitors, thus safeguarding reputation of quality.
Data Collection and Data Monitoring	Collect manufacturing cycle time and all other manufacturing costs accurately and efficiently. Monitor performance accurately and efficiently. Improve methods for the preparation of quotations through historical information. Reduce machine down-time while waiting for inspection of first-off. Establish accurate standard times.
Rapid Response Facility	Promote information sharing between customer and suppliers. Reduce production lead-times to less than that of competitors. Reduce prototyping lead-times to less than that of competitors.
Information Gathering	Promote information sharing between customer and suppliers.
DNC File Management	Improve small batch handling through reduction of programming prove-out time.
Inspection Audit and Control	Accommodate customer quality requirements in an efficient and cost-effective way. Improve quality standards above those of competitors, thus safeguarding reputation of quality. Reduce machine down-time while waiting for inspection of first-off.
Tooling Management	Provide information to minimize time waste.
Job Costing	Calculate cost implications for splitting and joining of batches. Collect manufacturing cycle time and all other manufacturing costs accurately and efficiently. Improve methods for the preparation of quotations through historical information.
Preventative Maintenance	Create stability.
Software Integration	Promote system integration within the organization. Promote system integration with all customers. Collect manufacturing cycle time and all other manufacturing costs accurately and efficiently.
Machine Tool Performance Monitoring	Establish accurate standard times. Eliminate time wasting. Improve small batch handling through setup time reduction. Collect manufacturing cycle time and all other manufacturing costs accurately and efficiently.
Delivery Monitoring	Improve Delivery reliability and predictability. Establish accurate standard times.

Table 1.7 Example of system evaluation against key requirements

Key MIS requirements	Mori Seiki MSC 518	Dialogue Dlog	ERT Seiki	GNT DNC Max	Alta Systems Real Vision	Tech. Systems
Shop floor information display	4	4	4		4	4
Shop floor data collection		4	4			4
...	...	...	...	...	...	...
Other features						
Editing facility	4	4	4	4	4	4
Photographs displayed		4	4	4	4	
.	...	...	...	...	...	...

The key MIS requirements list proved extremely valuable in providing guidance to the design and implementation of this system. In fact, the MIS was designed and developed in such a way that each of the twelve requirements was cross-checked. This cross-checking ensured that relevant modules and functions were built into the system, and that all the requirements would be satisfactorily supported. The following provides an overview of the system structure, and examples to illustrate how some of the key requirements were supported by the system.

System structure

The analysis as outlined above helped the company to develop its MIS system, with the overall objectives:

- To set up a direct data link via modem, so that drawing files from a customer's CAD system could be transmitted into the company's CAM system without the need to edit or reconstruct drawing elements.
- To allow the transmitted CAD drawing elements to be used to generate cutter paths ready for post-processing to any suitable and available CNC machine tool.
- To cut prototyping lead-times, both by reducing CNC programming time and by reducing the time for CNC program verification at the prove-out stage.
- To provide machine operators with job-related information in a focused and user-friendly manner.

Essentially the MIS system evolved from the integration and utilization of stand-alone software that was already being used in the everyday operation of the company. The fundamental essence of the system was to bring together existing and new software in an integrated way, resulting in the gathering and distribution of essential data, and the satisfaction of the key MIS requirements. The overall system structure is shown in Figure 1.30. This figure shows the company database with the proprietary software's scheduling, CAM and CAD systems all supplying data to the MIS system. In addition, photographic information is supplied as a visual aid in the system. The gathering of shop floor information, including machine tool monitoring and the time spent by operators on each job, are fed back into the MIS system.

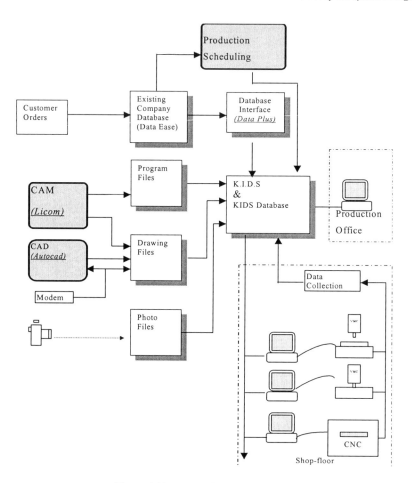

Figure 1.30 Layout of the MIS system

Management and utilization of plant and resources

This section illustrates the system's ability to satisfy some of the key requirements under this heading. For example, when first deciding on the way in which information should be accessed and displayed, it was considered important that the user found the system easy to operate and understand. In addition, the system had to provide assessable, relevant information to the task at hand. It was hoped that the user would have more incentive to use the new system if it provided useful information in a logical and efficient way. Traditionally, the case company and most other manufacturers of machined mechanical components have issued job cards/route cards, as detailed as necessary, with each batch of components launched on the shop floor. Within the case company, this paper document had evolved from carrying basic instruction for what were essentially straightforward jobs (e.g., "rough and finish turn complete"), to providing more sophisticated information. It was decided that the MIS would mimic some of the traditional approaches, both in operation and in visual presentation, This would allow the operator of the system to feel immediately at home and be able to relate to the proposed MIS system. By adopting this approach, the traditional job card was used as the front menu for obtaining focused, task-centered information required to satisfy the management and utilization of plant and resources. Hence, the system was designed to provide the following information:

- Job cards—manufacturing documentation.
- CAM information—cutter paths, feeds and speeds.
- Photographs—component and fixture recognition.
- Drawings—stage manufacturing drawings and final drawings.
- Scheduling information—machine work-to-lists and forward visibility.
- Machine tool information—capacity, achievable tolerances.
- Tooling information—tools required, cutter life, feeds and speeds.
- Part programs—proven or unproven files, recent edits.

The component job card, taken from the database, acted as the menu for the selection and displaying of information. This simple approach to information selection via the job card was readily accepted by all users and allowed the system to evolve when information from other sources was integrated.

Management and utilization of shop floor data

Four of the key MIS requirements listed under this heading were Data Collection and Data Monitoring, Delivery Performance Monitoring, Machine Tool Monitoring and Job Costing. All of these key MIS requirements relied on receiving information from the shop floor. Receiving accurate information from the shop floor was equally as important as providing accurate information to the shop floor. It could be argued that receiving false form information from the shop floor by way of collected data could be more detrimental to the overall manufacturing function than supplying inadequate information. This was because false information received could lull the operator into a false sense of security. Consequently, shop floor data collection and monitoring was designated as a key MIS requirement.

In particular, Delivery Performance Measuring was seen as the overall measure of Delivery reliability within the company. The seven companies that participated in the customer survey each monitored their suppliers in different ways. At one extreme, some customers appeared not to be monitoring their suppliers at all, and at the other extreme, some customers had fairly complex ways in which they measured delivery performance, the results of which were taken seriously. In most cases, information required for delivery performance measuring could be obtained from the company database, since information such as date of order placement, due date and customer date delivered were readily available for every job. However, in one case, the way in which the customer's suppliers were officially monitored was complex, involving additional information to be retrieved from the database. At this stage, the only information available on the system would be concerned with delivery performance. This information was obtained from the company database and entered into the Microsoft Jet Engine database where delivery monitoring parameters specific to each customer were displayed.

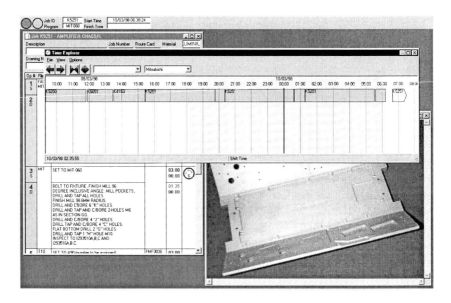

Figure 1.31 Visual display of machine tool monitoring

As far as machine operational data are concerned, the system collected data in real time and displayed machine tool cycle times in the form of a Gantt chart. The display also contained the relevant job card and, if necessary, a photograph of the component being machined (Figure 1.31). The Gantt chart could be seen for one particular machining center, and the cycle time length of three different pallets was displayed. This display could be called up on any of the workstations, either on or away from the shop floor. A machine tool could be selected and monitored to see if the machine tool was operating, and operating times compared with the standard times that had been set. It was also possible to check the same information from a remote location using a modem.

A related requirement to the above was job costing. The ability to be able to calculate the cost for manufacture of a component is paramount in the subcontracting manufacturing environment. A system for the initial cost estimation that is accurate, consistent, effective and quick is important when dealing in a competitive market environment. Equally, to be able to efficiently collect the data necessary to be able to accurately calculate the true manufacture is important. Job costing, which encompasses both the initial estimation of the cost of component manufacture and the calculation of the actual cost of manufacture upon completion, was identified as a key MIS requirement for the company. The costing system was designed to enable the user to retrieve historical data from the company database. This could include past job cards of manufactured components identifying the equipment used at that time, together with the standard time and actual time taken for each operation. This, together with stored photograph and drawing files (when available), enabled the user to employ the system as a historical reference. This ability proved extremely useful for cost estimation of similar components. Manufacturing instructions for all produced parts were broken down into individual operations. When completed, these instructions were stored/archived and could be recalled to reveal the associated cost of each individual operation calculated. This was particularly useful for the cost estimation of new parts that had similar features or characteristics to parts machined in the past, as shown in Figure 1.32.

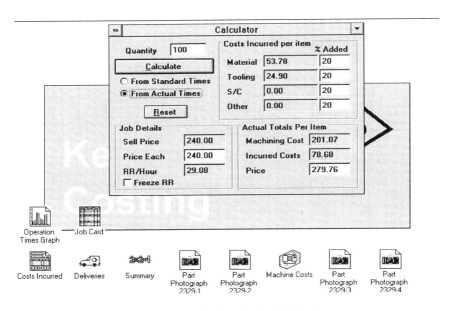

Figure 1.32 Systems display of costing/calculation menu

Additional system enhancements

A particularly important strategic requirement was the ability to provide a rapid response facility for prototyping services. With time-to-market pressures, early design of component parts are need for evaluation. Typically, in the early stages of development small quantities of parts, sometimes only one-off, are urgently required for evaluation before proceeding with the next development stage. The pressure is on for the designer to produce a drawing of the part as quickly as possible and for the manufacturer to make it as quickly as possible.

The system handles the rapid response information transmitted from customers through a process called "information chain." The customer uses the Internet to provide three-dimensional CAD files, in IGES format, of the component part required by rapid response. The file is viewed on the company CAD, and price and delivery is given to the customer. If necessary, costing would have been used for this purpose. Once a price and delivery is agreed, the relevant drawing file is copied from the CAD system to the CAM system. At this stage, material is obtained and, if necessary, the CAD file is plotted. Because predefined parameters have already been set, all drawing tolerances are known, together with material specifications and surface finishes, etc. The relevant profiles are captured within the CAM system, and cutter paths are simulated. A tooling list is automatically generated within the CAM database, and identification numbers assigned. Once the CAM user is happy with the cutter path simulation, the CAM file is post-processed for the designated machine tool on which the component will be manufactured. Concurrently, customer order details are entered into the company database and a production engineer writes the component job card, which is identified as a rapid response job.

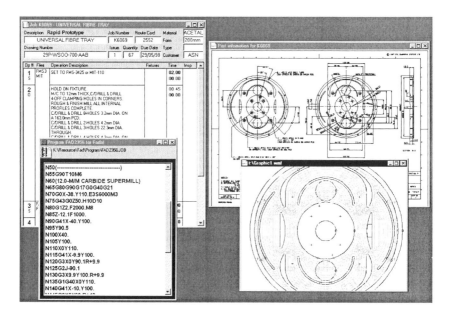

Figure 1.33 Information and display (rapid response facility)

The production engineer decides how the component will be manufactured, assigning the number of operations and machine tools to be used, and estimating the standard time for each operation. If the appropriate machine tool is available, the machine tool operator can be alerted and the system interrogated to find the rapid response job card. At this stage, the system should contain a detailed manufacturing description (job card), the customer's drawing, a tooling list, a cutter path simulation, and the part program file, which has been identified as an unproven file. By using these facilities and by working closely with customers, manufacturing lead-times can be reduced significantly, thereby playing an important part in helping customers to reduce the time taken for their designs to reach the market place. Figure 1.33 shows a typical component that has been manufactured under the rapid response facility. The figure includes a graphical display of a cutter, a cutter path, the job card, the part program file and the customer's drawing of the component.

Table 1.8 Component life cycle and information gathering

Customer Component Life cycle	Typical Batch Size	Customer Response/ Requirements	Manufacturer's Response/ Requirements	KIDS User Interface/Display
Prototype	1	CAD file	Rapid response Value engineer	Display prototype job card Display cutter paths Display prototype drawing Display initial tool list
Certification	3	Revised CAD file	Quick response	Display revised job card Display revised cutter paths Display drawing Display revised tool list Display photograph of part
Pre- production	10	Revised CAD file	Refine manufacturing methods	Display revised job card Display revised cutter paths Display revised drawing Display revised tool list Display photograph of part Display fixture photograph
Production	20	Cost justification	Optimize manufacturing methods	Display optimized job card Display optimized cutter paths Display drawing Display optimized tool list Display photograph of part Display fixture photograph Display stage drawings Display critical dimensions
Increased Production	50	Decrease cost	Additional optimization	As above, plus: Display fixture set up Information on production problems Inspection history
Decreased Production	20	Maintain cost	Reduce set up times	As above plus any optimizations made during full production
Spares	5	Reluctant price increases, no manufacturin details	Recall manufacturing methodology	All past information held within KIDS

When a component is first manufactured using the rapid response facility, information is gathered in the form of CAD files or drawings from customers, from which a job card and other details are written. The same is true if the component is manufactured under normal conditions. With components that start as development components, it is hoped that pre-production and production runs will follow. It is recognized that as the product matures, and with the experience of various production runs, continuous improvements to manufacturing techniques can be introduced. In order to do so, however, information needs to be gathered and refined as components pass through their respective life cycles. Table 1.8 shows typical information-gathering and displays the various stages of a customer's component life cycle on the system.

Case observations

Demands on the manufacturing industry to provide quality, flexibility and cost reduction have put pressures on manufacturing companies to improve productivity. These demands, coupled with computer hardware and software advances, have encouraged MIS development. Consequently, the role and importance of MIS within the manufacturing environment have changed dramatically in recent years. However, the initial design of such a system must be very carefully considered. The way in which it is structured and organized will have a profound effect on the way in which information can be delivered and utilized to support the company's strategic aims. This case study has attempted to address the key question of how to link the strategic and MIS requirements logically. The application of the proposed approach has helped the case company to develop an integrated system to support its strategic intentions which, in turn, has enabled the company to:

- Improve prototyping quality and lead-time by downloading engineering information directly from the customer's CAD system. This information is then used to generate cutter paths ready for post-processing.
- Improve cost control by providing online data collection and real-time analysis.
- Increase operational efficiency by providing operators with job-related information in a focused and user-friendly manner.

Through an analysis of the company's strategic manufacturing requirements, the proposed procedures revealed a number of MIS related issues and features that would help to ensure a competitive edge. A total of twelve key MIS requirements were established. These proved to be extremely valuable in providing guidance to the design and implementation of its MIS system, providing cross-checking between MIS functionality and the company's future strategic requirements. The resulting system has been seen as an effective "manufacturing strategic driver" to help this company maintain its competitive edge by improving part prototyping quality and lead-time, improving cost control through online data collection and real-time analysis, and increasing operational efficiency through with job-related information. Due to its success, the system was given the *UK Machinery Award for Innovation in Production Engineering*, for being "the most innovative application of computer technology in the manufacturing environment."

1.6 CONCLUSION

In facing the challenge of modern manufacturing, successful companies need skilled professionals and effective tools to design and manage world-class manufacturing and supply systems. A logical MSM framework helps to set systems thinking into the context of manufacturing systems management. This is defined as a domain that involves the necessary activities needed to regulate and optimize a manufacturing system as it progresses through its life cycle. Providing logical guidance for a company's MSM activities, its structure and contents help achieve understanding of the problem domain, and provide a basis for the development and adaptation of effective approaches and tools in practice.

This chapter has outlined the main functional areas, specified the generic processes and contents of these areas, and then integrated them into a closed loop to provide the basis for the development of a set of coherent processes and tools, and a means of bridging the existing MSA/MSD/MSO gap. Within the system design area, in particular, the framework also provides a design process reference architecture structured to support systems engineering principles. From the perspective of a system's life cycle, the MSM reference structure provides a more complete framework to link manufacturing strategy and a system's specifications. It not only provides the conceptual structure and sequence of the design process, but the means of describing the system itself. The cases of its industrial application have clearly demonstrated its practical value. For example, the greenfield MSD project has effectively used the approach to design and implement all MS processes required for the new factory in time for its operation and in line with the strategic targets of the organization. In addition to highlighting the need for the structured approach, the key learning points of these cases include the strategically-driven and simultaneous engineering approach that must be applied in process design and process ownership.

The complete, task-centered MSM workbook will be presented in the following chapters:

- *Chapter 2 Manufacturing and supply strategy analysis*. This chapter provides a set of task documents to help analyze, capture and/or develop future MS strategy.
- *Chapter 3 MSA/MSD interfacing*. This chapter provides a set of task documents to help link MS strategic requirements to MSD actions.
- *Chapter 4 MSD task execution*. This chapter presents the key principles and techniques involved in the execution of MSD tasks. It also provides a selection of generic MS design task documents, as well as a set of worksheets to help achieve the complete specification of an MS system.
- *Chapter 5 MS system implementation*. This chapter provides a selection of generic MS task documents related to system implementation, through which relevant techniques, such as those of project and change management, are logically incorporated into the MSM framework.
- *Chapter 6 MS system performance measurement and status monitoring*. This provides a set of task-documents related to the setting of project objectives, targets and constraints. In addition, the task documents of system status monitoring complete the MSM loop (strategy analysis—system design—system implementation—system status monitoring—strategy analysis).

Finally, issues related to the MSM framework's institutionalization within an MS organization and its further application in practice will be discussed in Chapter 7.

CHAPTER TWO

Manufacturing and Supply Strategy Analysis

2.1 INTRODUCTION

Following the structure of the MSM framework as presented in the previous chapter, this and the following chapters present a self-contained workbook. Using a task-centered approach, this workbook aims to guide the user step-by-step through the complete cycle of MS strategic analysis, MS system design and MS system status monitoring.

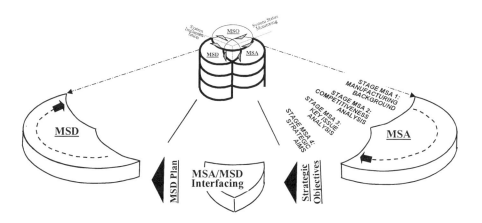

Figure 2.1 MS strategy analysis within MSM

This chapter focuses predominantly on the MSA (*manufacturing and supply strategy analysis*) process within the MSM domain. It provides a set of task documents to help capture a company's current MS strategy and its supporting information, and/or develop the organization's future strategic direction. Initially, an outline of the overall approach will be presented. Then a more detailed description of the tasks and processes involved in each of these stages will be given. An example will be provided to illustrate the steps involved.

The procedures are primarily directed at the formulation of MS strategic initiatives to guide the subsequent MSD projects. The overall structure of the process is as shown in Figure 2.1. As can be seen, the MSA section consists of four main stages, each of which comprise a number of task documents with a series of questions and methods of data collection:

- **Stage MSA 1—Manufacturing Background.** This provides a means of classifying the current state of development of the MS system and the role of the manufacturing function within the organization. It consists of a series of questions relating to the organization and the manufacturing system. These help to identify the requirements of the MS system and to define appropriate Product groups.
- **Stage MSA 2—Competitive Advantage.** This stage aims to capture data related to the marketing requirements and manufacturing performance for each of the Product groups. Competitive criteria are specified, order winners and qualifiers are identified and the results of the analysis are profiled. This determines the areas of the enterprise in which the organization needs to focus its allocation of resources, prioritization of activities and initiatives. Based on these, key success factors can be identified for the markets in which the enterprise is operating. The MS function must contribute accordingly in order to attain a competitive business position.
- **Stage MSA 3—Key Issues.** This stage starts with a gap analysis of the requirements and performance of the Product groups. From this, an initial indication of strategic requirements can be derived. This is followed by a SWOT (*strengths, weaknesses, opportunities* and *threats*) analysis of the Product groups. The results are then used to define the key issues and initial strategic objectives.
- **Stage MSA 4—Strategic Aims.** This stage aims to specify the details of the organization's future MS strategy. If a current strategy already exists, then it can be captured through a series of questions. Next, its principal policies are assessed with respect to the competitive criteria. The future policy can then evolve from the existing strategy, and the strategic aims can be derived from the key issues.

2.2 STAGE MSA 1—BACKGROUND ANALYSIS

This section involves gathering the relevant background and environmental information. This is done by classifying the current state of development of the MS system and the role of its function within the organization. Such information will provide indication about the relationships between the MS organization and its operation, and between the relevant functional strategies and the enterprise's business and corporate strategies. Ideally, the business strategy should be available for the analysis, together with relevant elements of the organization's technology, product and market strategies. The analysis process consists of a series of questions related to the organization and the MS system that need to be answered through the tasks shown in Figure 2.2:

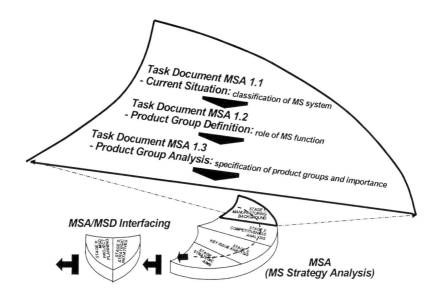

Figure 2.2 Stage MSA 1—MS background analysis

- Task Document MSA 1.1—*Current situation definition/classification*. This task aims to obtain an understanding of the state of the MS system within the overall context of the organization.
- Task Document MSA 1.2—*Product group definition*. This task captures data concerning the company's products, analyses them and then classes the products into logical Product groups.
- Task Document MSA 1.3—*Product group analysis*. Against a set of relevant parameters, this task conducts an assessment of the Relative importance of the groups with respect to their contribution to the performance of the business.

Just like a journey-planning exercise, the aim is to answer the question: *where are we now*? The completion of the related worksheets is a straightforward process of responding to a number of questions and completing a series of tables. This produces the following results:
1) classification of the business and its MS system,
2) definition of the role of the MS function,
3) specification of products and Product groups, and
4) identification of Relative importance of Product groups.

Table 2.1 Example—results of stage MSA 1.3

Products	A	B	C	D	Service A	Service B
Volume/ yr	23,000 ton	1000	1000	30,000	30,000 ton	?
Sales	$13.5 M	$4 M	$3.73 M	$5.58 M	$180,000	$260,000
% Sales	50.1%	14.5%	13.5%	20.3%	0.7%	0.9%
% Contrib'n	21.1%	12.3%	28.8%	34.4%	1.1%	2%
Market share	12%	30%	35%	35%	2%	2%
Growth	Very Good	Very Good	Very Good	Good	Good	Excellent
Innovation (out of 10)	Low (2)	Low (3)	Medium (6)	Low (3)	Low (2)	Medium (5)
Life cycle	Mature	Mature	Mature	Mature	Mature	N/A
Principle Processes	Slitting ERW	Machining Assembly	Machining Assembly	Threading & Painting	Shear cutters	Machining
Profit/sales	5%	10%	25%	20%	15-20%	25%
Typical order size	100 to 2000	No typical size	No typical size	Minimum 50	Use excess capacity	None
Market	Agriculture & industrial	Agriculture	Agriculture	Agriculture	Industrial	Industrial
Importance	20%	12%	30%	35%	1%	2%

For example, a company produces one main type of products, and undertakes a number of subcontracting roles. It has one key customer, who sells on the products to the end-users and several smaller customers. The business can be considered to be a small-to-medium sized enterprise. Its manufacturing system is largely batch manufacture. The process is based on traditional machine and assembly shops, and operates cellular manufacture based on components rather than part families. The system structure is "make-to-order" from stock and from suppliers, with elements of assemble-to-order. The organizational structure has five levels, from director to operator, and is based on a functional focus. The company is about to undertake a brownfield reorganization for improvement. There are four Product groups, and two additional services, as shown in Table 2.1. Product group A represents regular, relatively high volume standard products, competing largely on cost, quality and Delivery lead-time. Group B represents a standard product with a number of variants, and competes primarily on cost, Delivery lead-time and Delivery reliability. Group C is similar to B but has a reduced number of variants and competes largely on quality, cost and Delivery lead-time. Group D is a relatively high volume product with a small number of variants, and competes on cost, Delivery lead-time and quality and has a similar market to A. The two service groups are very different. Service group A is relatively high volume, but uses only excess machine capacity and competes largely on Delivery lead-time, Delivery reliability and Volume flexibility. Service group B represents a non-core activity, manufacturing low volume, customized products that mainly compete on quality, cost and Design flexibility. This example will be used for illustration in the remainder of this text.

Task Document MSA 1.1—Current Situation Definition

TASK OVERVIEW

DESCRIPTION

This task provides a means of assessing the business, its organization and its manufacturing and supply system. The information gathered should assist management and/or the design team to come to a common understanding of the business from a corporate as well as a functional perspective, and should help define the role of manufacturing and supply within the enterprise. The completion of this task will clarify three issues: classification of the business and its organization, classification of the MS system, and a statement of the role of the MS function.

TASK INSTRUCTION

There are several reasons for classifying an organization and its MS system. These include generalizing operating conditions, transferring approaches from conventional to advanced systems and prescribing the most likely strategies and systems to succeed. The main aim is to allow the managers and system designers to formalize and increase their understanding of the business and the manufacturing function, and to provide a record of the state of the enterprise when the manufacturing strategy is proposed, and the system designed. This helps to achieve an understanding of the needs of the MS function, and its readiness for change. A number of classification techniques and taxonomies are available for enterprise and MS systems. For instance, the business can be classified with respect to its structure, culture and organizational behavior:

Business Structure: structural configuration, coordinating mechanism, key organizational section, decentralization type.

Business Culture: cultural orientation, organizational activities.

Organizational Behavior: growth, market, product development, new products and services, production, investment, concentration, cooperation, behavior towards competitors.

Similarly, the manufacturing and supply system can be codified with respect to its structure, including the product/process matrix, operating system, system relationships, system evolution and state, and system life cycle:

System Structure: product process matrix (volume, variety, system type, degree of technology integration, degree of technology automation, scale of capacity increment), stock and order operating system structure.

System Relationships: nature of business, customer influence, and organizational structure.

System State: degree and state of evolution.

System Life cycle: life cycle stage.

The definition of the role of the manufacturing function requires the management/design team to reflect on their responses to the above, to question the role of the manufacturing function in the organization and to specify the purpose of the manufacturing function.

Generally, this requests a textural statement of the manufacturing function definition for future reference. While the input of such a definition is not absolutely necessary, especially if there already exists a strategy document, this stage provides an opportunity to examine and assess the statement at an early stage. Its principal aim is to ensure that the strategy development group and the MSD designers have a common understanding of the role of the manufacturing function.

Similar background information can also be collected for the supply aspects of the organization, if so required.

TASK LINKS   POSITION IN MSM FRAMEWORK

INPUT FROM :	Cooperative strategies: Market Product Business	OUTPUT TO :	Task MSA 1.2

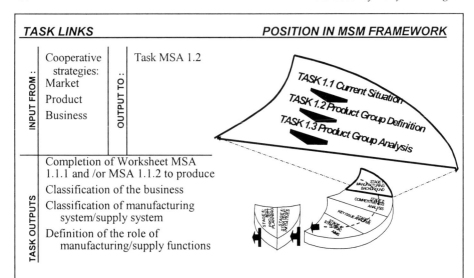

TASK OUTPUTS

Completion of Worksheet MSA 1.1.1 and /or MSA 1.1.2 to produce

Classification of the business

Classification of manufacturing system/supply system

Definition of the role of manufacturing/supply functions

TASK PROCEDURE   TASK FLOWCHART

	Input	Tool	Output
Step 1	Data to be collected as specified	Wksheet MSA 1.1	As specified by wksheet
Step 2	Data to be collected as specified	Wksheet MSA 1.1	As specified by wksheet
Step 3	Data to be collected as specified	Wksheet MSA 1.1	As specified by wksheet
Step 4	Data to be collected as specified	Wksheet MSA 1.2	As specified by wksheet

Collect business background information

Collect organisation background information

Collect manufacturing background information

Supply system relevant ? — Yes → Collect supply background information

No

WORKSHEET MSA 1.1.1—Manufacturing Background

Project Title:

Person(s) Responsible:

Version: **Date Completed:**

Business/organization classification
Business definition (*business—customers—competitors*)

Business structure
Structural configuration (*tree diagram*):

Configuration (*simple—machine bureaucracy—professional bureaucracy—divisionalized—Other*):

Coordinating mechanism (*direct—standard work—standard skills—standard outputs—adjustment – other*):

Key part of organization (*strategic apex—techno structure—operating core—middle line – support staff – other*):

Type of structure (*centralized—distributed—other*):

Size of company:

Business culture
Ownership:

Dominant culture (*power—role—task—person—other*):

Control and power within organization:

Organizational behavior

Organizational orientation (*entrepreneurial—bureaucracy—job/project oriented—person oriented—other*):

Strategic behavior (*growth—market—product develop—new products—production—investment—concentrate—cooperate—compete*):

Operating environment

Business purpose:

Prevalent technology:

MS system classification

System structure

Product/process matrix:

Process type and role of manufacturing, operating system structure (*make-to-stock, make-from-stock, make-to-stock from supplier, make-to-order from stock, make-to-order from supplier*):

System relationships

Customer influence on manufacturing, Organization (*hierarchy—functional—matrix—product focus—temporary—other*):

Organizational structure (*tree diagram*):

System state

Evolution (*complex—simple—integrated—automated—computerized*):

System life cycle (*greenfield—growth—maturity—improvements—brownfield—maturity—improvements—decline*):

WORKSHEET MSA 1.1.2—*Supply Chain Background*

Project Title:

Person(s) Responsible:

Version: **Date Completed:**

External factors

Economic issues

Inflation rate:	
Currency conversion rates:	
Economic growth rate:	
Employment rate:	
GDP/ GNP:	

Regulatory issues

Product standards and specifications:	
Labor laws:	
Commerce laws:	
Social regulations:	
Health & safety requirements:	
Environmental aspects:	
Product & process recycleability:	
Usage of clean technology:	
Emission controls:	
Conformance to regulatory laws:	

Technology

Information technology advancements (general):
Telecommunications advancements:
Enterprise resource planning systems and techniques:
Handling and shipping equipment and tools:

Globalization

Trade barriers:
Cross border restrictions:
Restrictions in export/import operations:
Emerging markets:
Affiliate markets:
Alliances/mergers/spin off:

Internal factors

Types of facilities

Manufacturing units:	
Warehouses:	
Depots:	
Transshipment points:	

Number of facilities

Manufacturing units:	
Warehouses:	
Depots:	
Transshipment points:	

Size/capacity of facilities

Manufacturing units:	
Warehouses:	
Depots:	
Transshipment points:	

Type

Manufacturing units type	

Warehouses type:
    Market oriented
    Manufacturing oriented
    Intermediate	

Distribution channels

	No. 1	No. 2	No. 3	….
Distribution channels				
Distribution lines				
Distribution points				

Transportation

Modes of transportation

	No. 1	No. 2	No. 3	….
Types of vehicles				
Number of vehicles				
Cost				

Shipment size

	No. 1	No. 2	No. 3	….
Volume of shipment per vehicle				
Shipment time				
Cost				

Task Document MSA 1.2—Product group Definition

TASK OVERVIEW

DESCRIPTION

A thorough understanding of the company's products and market is essential when developing an MS strategy and when executing an MSD project. A company can often be viewed as being composed of a set of product segments, and these product segments must compete in the market in their specific ways. The definition and analysis of Product groups is therefore an important process. It provides the basis for the subsequent analysis.

TASK INSTRUCTION

The task of Product group definition uses a number of simple tools to aid the specification of Product groups or families. First, data regarding product types, variants, etc., should be collected. Quantitative and qualitative analysis can then be carried out to assist the selection of major Product groups based on similarities of features. Typical criteria may include: volume, variety, costs, profits, markets, typical customers, competing criteria, life cycle stage, resources and principal processes, materials, typical order size, degree of standardization, and product introduction rate. While there are many ways of forming Product groups, the methods recommended here are based on the identification of products which share similar competitive criteria, similar product life cycle stages, and/or are based on the result of product volume-variety analysis. To assist in this process, a number of techniques can be applied for identifying runners, repeaters, and strangers, as shown in the figure below. To a large extent, this product/market-based method relies on user knowledge. Worksheet MSA 1.2.1 requests and presents the information in a manner that makes the tasks more structured. The results from this task should be a series of clearly defined Product groups.

Once each of the product ranges is defined diagrammatically, the table of Worksheet MSA 1.2.1 provides a number of variables to help divide products into groups. It is probably easiest to consider a product family as a grouping of products which compete in the market in similar ways. Where market segments are being applied, it should be noted that products in the same segment may win orders in different ways. Other useful indicators include the product life cycle concept and the manufacturing operations and production processes that are required for the constituent parts.

During this step and the subsequent analysis step, it is useful to have available copies of the product and marketing strategies, if they exist. Following a similar approach, products can also be grouped according to their supply characteristics, if so required (the purpose of Worksheet MSA 1.2.2 is the same as that of Worksheet MSA 1.2.1, but with additional items specifically relevant to supply chain analysis).

This figure shows how data can be collected and analyzed for product-volume analysis.

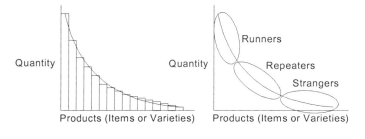

TASK LINKS   POSITION IN MSM FRAMEWORK

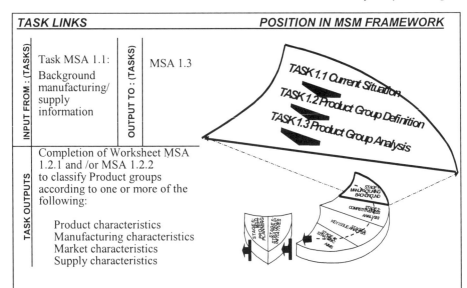

INPUT FROM : (TASKS)
Task MSA 1.1:
Background
manufacturing/
supply
information

OUTPUT TO : (TASKS)
MSA 1.3

TASK OUTPUTS
Completion of Worksheet MSA
1.2.1 and /or MSA 1.2.2
to classify Product groups
according to one or more of the
following:

    Product characteristics
    Manufacturing characteristics
    Market characteristics
    Supply characteristics

TASK PROCEDURE   TASK FLOWCHART

	Input	Tool	Output
Step 1	Initial data regarding product types and structures	Worksheet MSA 1.2.1/1.2.2	Initial product family hierarchies (tree diagrams)
Step 2	Data to be collected as specified	Worksheet MSA 1.2.1/1.2.2	
Step 3		Product – Volume analysis	
Step 4		Assign Product groups	Completion of Worksheet MSA 1.2.1

Flowchart:

Initial definition of product groups

↓

Product data selection

↓

Product/Volume analysis → Product groups satisfactory ?

No / Yes

Assign product groups

WORKSHEET MSA 1.2.1—Product group Definition

Project Title:

Person(s) Responsible:

Version: **Date Completed:**

Product family hierarchies (tree diagram)

Product analysis

Products						
Variants						
Volume						
Life cycle stage						
Principle processes						
Materials						
Approx. profit/cost/sales						
Order size						
Standardization						
Product intro. rate						
Market						
Competing criteria—quality						
Competing criteria—Delivery lead-time						
Competing criteria—Delivery reliability						
Competing criteria—Design flexibility						
Competing criteria—Volume flexibility						
Competing criteria—cost/price						
Customers						
Other						

WORKSHEET MSA 1.2.2—*Product Factors for Supply Analysis*

Project Title:

Person(s) Responsible:

Version: **Date Completed:**

Load units/function units (for each Product group)

Product group	A	B	C	D	E	...
Number of units per pellet						
Number of units per tote box						
Cycle stock levels						
Inventory levels						
Safety stock levels						

Material handling equipment (per Product group/facility)

	Site 1	Site 2	Site 3	Site 4	....	Site N
Types of equipment						
Pick rates						
Equipment utilization						
Level of automation						
Lines/man hour						
Space utilization						

Personnel costs

Personnel costs	Site 1	Site 2	Site 3	...	N	Total
Salaries						
Personnel cost (% of distribution cost)						

Operation costs

Energy						
Depreciation						
Taxes						
Rent						
Maintenance						
Communications						
Material costs						
Operation cost (% of distribution cost)						

Inventory costs

As % of turnover						
As % of turn rate						
Carrying cost (% of total inventory)						
As inventory carrying cost (% of sales)						

Asset conditions

Payback period						
Net present value (NPV)						
Rate of return						
Operation costs						
Maintenance costs						

Task Document MSA 1.3—Product group Analysis

TASK OVERVIEW

DESCRIPTION

This task provides an in-depth analysis of the previously defined Product groups. By comparing them through the use of relevant criteria, a measure of Relative importance is specified for each with regard to the operation of the enterprise. Typically, these criteria include parameters such as sales, profit, volume and market share. Once defined, these measures of Product groups will be used frequently at later stages in the MSM framework. Each of the criteria should be assigned a relative ranking based on the company's assessment of its importance. Once all the Product groups have been assessed accordingly, a relative overall measure can be assigned to each group. When assigning the relative values of importance, care should be taken of factors and relationships within the operations, business and market environments including, for instance, the expected growth, products in different Product groups being supplied to the same customers and customer development strategies. The output is a series of tables detailing the various criteria for each Product group, and the overall relative ranking of the Product groups.

TASK LINKS

POSITION IN MSM FRAMEWORK

INPUT FROM : (TASKS)

MSA 1.1: Product & market strategies, background information.

MSA 1.2: Product group definition

OUTPUT TO : (TASKS)

MSA 2.1

TASK 1.1 Current Situation
TASK 1.2 Product Group Definition
TASK 1.3 Product Group Analysis

STAGE 0:
MANUFACTURING
BACKGROUND

STAGE 1:
COMPETITIVENESS
ANALYSIS

STAGE 4:
STRATEGY &
PLANNING

STAGE 3:
STRATEGIC
INITIATIVES

KEY ISSUE ANALYSIS

STAGE 2:
STRATEGIC
AIMS

OUTPUTS

Performance information of Product groups

Relative ranking of Product groups

TASK PROCEDURE

TASK FLOWCHART

	Input	Tool	Output
Step 1	Product & market strategies.	Worksheet MSA 1.3.1	List of relevant criteria to be used for analysis
Step 2	Product group definition	Worksheet MSA 1.3.1	Relevant importance value for chosen criteria
Step 3		Worksheet MSA 1.3.1	Performance data of group against criteria
Step 4		Worksheet MSA 1.3.1	Completion of Worksheet MSA 1.2.1

Choose relevant criteria

Specify relative importance for each criterion

Collect relevant data

Calculate relative importance of product groups

WORKSHEET MSA 1.3.1—Product group Analysis

Project Title:

Person(s) Responsible:

Version: **Date Completed:**

Product group		A	B	C	D	....
	Relative importance of criterion (*v*)					
Sales						
% Sales						
% Contribution						
Volume						
Market share						
Customers						
Competitors						
Product life cycle stage						
Product intro rate						
Growth opportunities						
Vulnerabilities						
Breadth of group						
Standardization						
Degree of innovation						
.....						

Relative importance of Product group (*I*)

The resulting importance criteria can be described as a vector *I*, being the product of variable importance vector *V* and a variable matrix *M*, where: $V = [\,v_1,\ v_2,\ ...,\ v_n\,]$, and v_i represents the importance of the ith individual Product group analysis variable. The variable matrix *M* is defined as shown (assuming four Product groups are involved: A, B, C and D), where, for example, b_i represents the ith Product group analysis variable for Product group B.

$$M = \begin{vmatrix} a_1 & b_1 & c_1 & d_1 \\ a_2 & b_2 & c_2 & d_2 \\ ...... \\ a_i & b_i & c_i & d_i \\ ...... \\ a_n & b_n & c_n & d_n \end{vmatrix}$$

The dimension of the matrix depends on the number of Product groups and the number of variables being considered.

In this example: $I = [\,I_a\ \ I_b\ \ I_c\ \ I_d\,]$

where, for example: $I_b = [v_1 \cdot b_1 + v_2 \cdot b_2 + ... + v_i \cdot b_i\ ... + v_n \cdot b_n]$

2.3 STAGE MSA 2—BASIS FOR COMPETITIVE REQUIREMENTS

The general aim of stage MSA 2 is to answer the question: *where are we now?* It is designed to capture the marketing requirements and the actual system performance in relation to each of the Product groups, and/or the system as a whole. This information enables a competitive requirement profile to be developed for each of the Product groups and for the whole system. It indicates the areas of the enterprise in which the organization must focus its effort in order to achieve a superior position in relation to its competitors. The four task documents involved in this stage, together with their overall outputs, are shown in Figure 2.3.

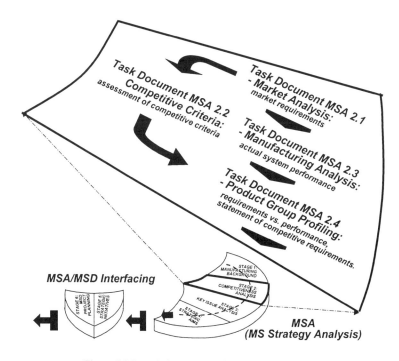

Figure 2.3 Stage MSA 2—Basis for competitive requirements

Tasks one and three are essentially concerned with data collection and analysis activities. The customer and market requirements are identified with respect to a number of six major competitive performance criteria. Secondary criteria can also be used, if deemed important. Similarly, the performance of the current manufacturing function is analyzed with respect to the same competitive criteria. Task two involves the subsequent derivation of the Order winning and qualifying criteria using the evidence presented in the market analysis. Finally, the fourth task captures the relevant information with respect to the six competitive criteria, and produces a number of requirements/performance profiles. It also produces a textural entry of the statement of the basis for the manufacturing function's competitive advantage.

The market requirements analysis of the example company's individual Product groups produces the results as given in Table 2.2, against the six competitive criteria used for the analysis.

Table 2.2 Example—summary of Product group requirement analysis (*Worksheet MSA 2.1.1*)

Requirements (0 – 100)	Group A	Group B	Group C	Group D	Service A	Service B
Quality	90	95	75	85	90	90
Delivery lead-time	70	90	90	90	80	80
Delivery reliability	60	90	90	90	70	85
Design flexibility	60	80	80	80	80	90
Volume flexibility	60	85	85	85	80	80
Cost	90	80	75	75	80	75

The definition of competitive criteria, using Worksheet MSA 2.2.1, reveals the information as outlined in Table 2.3.

Table 2.3 Example—summary of order winners and losers (*Worksheet MSA 2.2.1*)

Competitive criteria	Group A	Group B	Group C	Group D	Service A	Service B
Order winners	Lead-time, reliability	Cost	Volume flex., design flex.	Quality, volume flex.	Reliability, lead-time	Reliability, volume flex.
Order qualifiers	Cost, quality	Quality, lead-time, reliability	Lead-time, reliability	Lead-time, reliability	Quality	Quality, design flex.
Potential order winners	Lead-time	Volume flex.	Quality		Volume flex.	Cost
Order losers	reliability		Cost		Lead-time	Reliability

In addition, the manufacturing analysis (Table 2.4) reveals that good levels of quality are being achieved. There are also acceptable levels of Design flexibility, Volume flexibility and cost, though these could still be improved.

Table 2.4 Example—summary of manufacturing performance analysis (*Worksheet MSA 2.3.1*)

Performance (0-100)	Group A	Group B	Group C	Group D	Service A	Service B
Quality	80	95	95	95	90	95
Delivery lead-time	60	45	55	70	90	85
Delivery reliability	60	50	65	60	95	95
Design flexibility	60	90	90	70	90	90
Volume flexibility	60	60	70	65	85	75
Cost	60	60	85	80	85	85

Task Document MSA 2.1—Market Analysis

TASK OVERVIEW

DESCRIPTION

It is necessary for a company to obtain an understanding of what is required of each of its products from the customers in the chosen market segments. These requirements should form the basis of all future investment in processes and infrastructure. The market analysis helps achieve this by investigating its business, markets, competitors, and the reasons why products are chosen by customers. A number of parameters are suggested as sample measures for each competitive criterion. Against these, the individual customer requirements are to be assessed for each Product group. The parameters are provided as guidelines and can be supplemented and customized to meet the specific requirements of the business, as shown in *Worksheet MSA 2.1.1*. This sheet is to be filled to develop an overall picture of the customer service requirements for all Product groups.

For the same purposes, *Worksheet MSA 2.1.2.* is to be used for the supply aspects of the operation. *Tool/Tech. MSA 2.1.2* provides a checklist of relevant measures to consider.

TASK LINKS

POSITION IN MSM FRAMEWORK

INPUT FROM :	OUTPUT TO :
Background Role of manuf. & supply Prod. group analysis	MSA 2.2 MSA 2.3

OUTPUTS

Market requirements in terms of : customer service demand on performance from the organization, measured against a number of competitive criteria on a Product group basis.

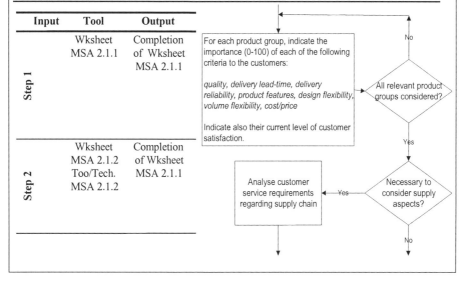

TASK PROCEDURE

TASK FLOWCHART

	Input	Tool	Output
Step 1		Wksheet MSA 2.1.1	Completion of Wksheet MSA 2.1.1
Step 2		Wksheet MSA 2.1.2 Too/Tech. MSA 2.1.2	Completion of Wksheet MSA 2.1.1

For each product group, indicate the importance (0-100) of each of the following criteria to the customers:

quality, delivery lead-time, delivery reliability, product features, design flexibility, volume flexibility, cost/price

Indicate also their current level of customer satisfaction.

All relevant product groups considered? No / Yes

Analyse customer service requirements regarding supply chain ←Yes— Necessary to consider supply aspects? No

WORKSHEET MSA 2.1.1 – Product Group Market Analysis

Project Title:

Person(s) Responsible:

Version: **Date Completed:**

Key Customer(s):

Product group / Importance						
Quality (0-100)						
Conformance to spec						
Reliability in use						
Customer satisfaction						
Delivery lead-time (0-100)						
Lead-time requirements						
Delivery change notice						
Customer satisfaction						
Delivery reliability (0-100)						
Delivery window						
Contractual Delivery lead-time						
Required Delivery lead-time						
Customer satisfaction						
Design flexibility (0-100)						
Design changes						
Customized products						
Customer satisfaction						
Volume flexibility (0-100)						
Minimum order size						
Maximum order size						
Average order size						
Seasonality demands						
One-off demands						
Predictability						
Order change notice						
Customer satisfaction						
Cost/price (0-100)						
Price sensitivity						
Margins						
Customer satisfaction						
Product features (0-100)						
Unique features						
Superior performance						
Customer satisfaction						
Other criteria						

WORKSHEET MSA 2.1.2—*Supply Customer Service Factors*

Project Title:

Person(s) Responsible:

Version: **Date Completed:**

Production group	A	B	C	E	...	Average
Order cycle time						
Entry						
Processing						
Pick & ship						
Transit time						
Consistency & reliability						
On time						
Inventory availability						
Product availability						
Part types (ABC)						
Order size constraints						
Order convenience						
Delivery time & flexibility						
Back order						
Expedite order						
Substitute order						
Transportation						
Invoicing procedures & accuracy						
Order completeness & accuracy						
Administration errors						
Order picking errors						
Shipping errors						
Claims procedures						
Complaints						
Claims						
Condition of goods						
Warehouse damage						
Company shipping damage						
Carrier shipping damage						
Quality						
Packaging convenience						
Sales service						
Product support						
Repair parts						
Repair service						
Technical advice						
Order status information						

TOOL/TECHNIQUE MSA 2.1.2—Measures Related to Supply Chain

Customer (dealer) requirement
Measures
Parts availability off dealer shelf as a %
Number of pieces per part stocked at the
dealer
% Split of lines required by dealers next
day
Dealer costs to achieve the desired parts
availability
Total supply chain logistic costs as % of
dealer net sales
Replenishment lead-times
Maximum order cycle time in hours
Emergency order cycle time in hours

Size-dealer inventory
Dealer stocking profiles
Immediate fill % dealer to customer
Replenishment fill (%)
Emergency/stock order ratio
Cycle time
Order types/frequency
Order volumes

Warehousing, inbound and outbound
transport performance measures
Number of order types———
Number/WSO (weekly stock orders)
Number/DSO (daily stock orders)
Stock order cycle
Pricing policy
Emergency orders————%
Surcharge or discount
Return allowance ———% on all parts;
Number of days return; % on fast
movers/% on slow movers
Number of PDC (parts distribution
centers) serving dealers

Transport system
Number of order lines
Transport costs per annum
Inbound labor cost
Distribution costs
Distribution costs
Number of load units

Number of functional units
Average walking time

Picking productivity
Material receipt to stocking time
Transcription effort: time & man-hours
Number of stock-outs
Picking costs, times and categorization
Multi-order picking
Batch picking
Single dealer picking
Stock keeping accuracy
Number of lines in the PDC

Throughput time
Number of suppliers
Volume throughput
Inventory size and cost
Delivery points
Part complexity
Immediate fill % dealer to customer
Replenishment fill (%)
Emergency/stock order ratio

Cycle time
Order types/frequency
Order volumes
Replenishment time
Supplier delivery frequency
% of deliveries via express transport
modes
Parcel service cost p.a. (per annum)
Parcel by air cost p.a.
Hub-dealer cost p.a.
CDC-hub cost p.a.
Inventory costs p.a.
Outbound labor costs p.a.
Inbound labor costs p.a.
Service strategies—- DSO/WSO; single
tier/multi-tier
Average revenue per dealer
Number of parts stocked
Number of parts stocked by category
Number of lines ordered by class –
Lines/day—DSO lines/day; VOR
lines/day
Safety stock levels

Task Document MSA 2.2—Competitiveness Analysis

TASK OVERVIEW

DESCRIPTION

This task aims to improve the understanding of the enterprise's markets and to identify the criteria by which each Product group wins orders. It requires an examination of the information derived from the market analysis in order to assess the Product groups with respect to the six competitive criteria: *Order winning criteria* (W). The factors that directly and significantly contribute towards how the company and its products win business. Customers are likely to look for a performance that is better than the competition; *Order qualifying criteria* (Q). The factors that, once above a certain level, permit the company and its products to be considered within the marketplace. Customers are likely to check that the product conforms and is within the range deemed acceptable in the market; *Potentially order winning criteria* (P). These are Order qualifying criteria with the potential to become Order winning criteria; *Order losing criteria* (L). These are Order qualifying criteria that are order-losing sensitive, such that a drop in performance resulting in lost orders.

The Product groups should ideally be assessed against the competitive criteria for the current period and for two time periods in the future. Typically such time periods would be three and seven years respectively, although they depend to a large extent on the industry in which the company operates. Reasons for any anticipated changes should be recorded. Similarly, where possible an analysis of the competitors' approaches should also be recorded and the reasons for any major differences noted.

TASK LINKS   POSITION IN MSM FRAMEWORK

INPUT FROM : | Task MSA 2.1   **OUTPUT TO :** | Task MSA 2.3

OUTPUTS | Classification of competitive criteria for Product groups

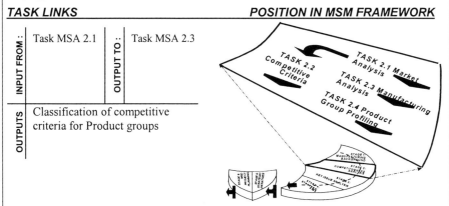

TASK PROCEDURE   TASK FLOWCHART

	Input	Tool	Output
Step 1	Data from market analysis, other data collection	Wksheet MSA 2.2.1	
Step 2	Data from market analysis, other data collection	Wksheet MSA 2.2.1	Completion of Wksheet MSA 2.2.1

For each product group, classify the competitive criteria (*quality, delivery lead-time, delivery reliability, product features, design flexibility, volume flexibility, cost/price*) according to:
 W: order winning
 L: order losing
 Q: order qualifying
 P: potentially order winning

All relevant product groups considered?   No

Repeat the previous for what to be expected in the next three and seven years respectively. Record reasons for change.

Yes

WORKSHEET MSA 2.2.1—Competitive Criteria Definition

Project Title:

Person(s) Responsible:

Version: **Date Completed:**

Identify: Order winning (W), Order losing (L), Order qualifying (Q), Potentially order winning (P)

Current period
Product/Product group

Quality

Delivery lead-time

Delivery reliability

Design flexibility

Volume flexibility

Cost/price

Other

Competitors approach
Product/Product group

Quality

Delivery lead-time

Delivery reliability

Design flexibility

Volume flexibility

Cost/price

Other

Own criteria expected after three years
Product/Product group

Quality

Delivery lead-time

Delivery reliability

Design flexibility

Volume flexibility

Cost/price

Other

Expected after seven years
Product/Product group

Quality

Delivery lead-time

Delivery reliability

Design flexibility

Volume flexibility

Cost/price

Reasons for change

Task Document MSA 2.3—Manufacturing Analysis

TASK OVERVIEW

DESCRIPTION

In contrast to what is required by the market, it is now necessary to obtain a detailed understanding of what is actually being achieved by the company regarding each of its Product groups, and the system as a whole, in the chosen market segments.

In a similar manner to the market requirement analysis, a number of parameters and performance indicators have been suggested for the manufacturing performance analysis (*Worksheet MSA 2.3.1*). For each overall heading, an indication of how well manufacturing is performing can be given on a scale from 0 to 100.

TASK LINKS   POSITION IN MSM FRAMEWORK

INPUT FROM.:

OUTPUT TO.: Task MSA 2.4

TASK OUTPUTS

An assessment of manufacturing performance in terms of what is actually being achieved by the company, based on the performance of Product groups against the competitive criteria.

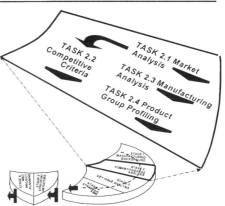

TASK PROCEDURE   TASK FLOWCHART

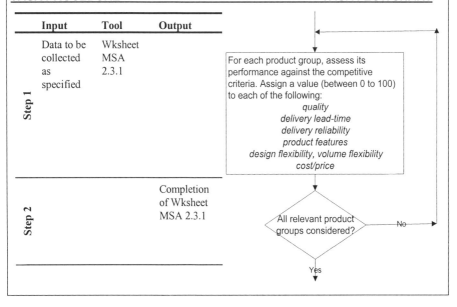

	Input	Tool	Output
Step 1	Data to be collected as specified	Wksheet MSA 2.3.1	
Step 2			Completion of Wksheet MSA 2.3.1

For each product group, assess its performance against the competitive criteria. Assign a value (between 0 to 100) to each of the following:
quality
delivery lead-time
delivery reliability
product features
design flexibility, volume flexibility
cost/price

All relevant product groups considered?   No

Yes

WORKSHEET MSA 2.3.1—*Product Manufacturing Analysis*					
Project Title:					
Person(s) Responsible:					
Version:		**Date Completed:**			
Product group					
Quality					
Actual quality level					
Customer reject rate					
Final failure rate					
Intermediate scrap rate					
Cost of scrap/cost of warranty					
Delivery lead-time					
Actual Delivery lead-time					
Manufacturing lead-time					
Schedule changeability					
Inventory investment					
Operation hours/total time in factory					
Delivery reliability					
Deliveries within window					
Complete orders					
Error-free orders					
Design flexibility					
Ability to cope with product range/change					
Design changes per year					
Ability to cope with design change					
Proportion customized					
Customization ability					
% increase in lead-time over std. product					
Volume flexibility					
Ability to respond to demand increase					
Product shelf life					
Minimum/maximum order size					
Setup times					
Seasonal/random demand variation					
Frequency of schedule changes					
Size of schedule changes					
Effect on Delivery lead-time					
Cost/price					
Actual cost incurred					
Manufacturing contribution: % sales					
Manufacturing contribution: /machine hr.					
Manufacturing contribution: /man hour					
Manufacturing contribution: overheads					
Manufacturing contribution: materials					
Manufacturing contribution: labor costs					
Non- manufacturing contributions					
Other criteria					

Task Document MSA 2.4—Product/System Profiling

TASK OVERVIEW

DESCRIPTION

Based on the results from the previous analysis of Product groups, this task constructs a series of utility values and profiles to assess the market requirements on the MS system, and the actual performance of the MS system in meeting those requirements. The aim of these profiles is to allow for a gap analysis to be executed (between the market demands and the actual system performance) within the MSA/MSD cycle, in order to identify areas needing improvement.

TASK INSTRUCTION

Following the technique of utility analysis, as described in *Tool/Technique MSA 2.4.1*, this task provides a structured approach to evaluate the effectiveness of the current manufacturing and supply operation:

1) Specify the Relative importance of each of the Product groups.
2) Identify Relative importance of each of the competitive criteria with respect to the Product groups.
3) Draw a Product group profile according to above.
4) Repeat the above, but attempt to establish the actual performance of the system.

A number of different requirements/performance profiles can be constructed for both the Product groups and the system as a whole, as described in detail in *Tools/Techniques MSA 2.4.2*. Together, these profiles provide a mechanism for both system-wide and product-group related method for evaluating MS requirements and MS performance.

The usefulness of different approaches and the comparison of approaches depends to a great extent upon the actual situation, the degree of focus of the facilities, the degree of complexity of the systems, the Relative importance of each Product group and the contribution towards the competitive criteria that each Product group provides. Gap analysis can be conducted in a flexible way depending on the needs:

1) Product-related requirement/system performance gap analysis. With this approach, the individual requirement profile of the key Product groups can be compared to the current system performance profile to identify the future strategic direction of the company. The manufacturing strategy thus developed will support the concept of the "focused factory" because the resulting system will be geared toward satisfying the manufacturing needs of the company's key products, and each product family becomes an individual manufacturing entity or unit. The competitive criteria can then be considered and optimized separately for each individual product family.

2) Factory-wide requirement/system gap analysis. With this approach, the overall requirement profile is compared against the overall system performance profile to identify the overall gap, formulating future manufacturing strategies which aim to satisfy system-wide manufacturing requirements. It should be remembered, however, that the construction of such utility functions is relatively simplistic (particularly the aggregated system profile) and, as such, they should be used with caution within the strategy analysis process. In effect, they essentially represent a compromise configuration for the manufacturing system. They should preferably be interpreted as an overview or a guideline of the requirements for the individual Product groups and for the system.

3) The maximum-specified-system gap analysis. A different means of using a system profile is to establish a weighted product profile, again based on the Relative importance of each of the Product groups. However, instead of accumulating and averaging these profiles, this approach constructs profiles by selecting the maximum requirement for each criteria.

TASK LINKS   POSITION IN MSM FRAMEWORK

INPUT FROM :	MSA 1.1, MSA 2.4	OUTPUT TO:	MSA 3.1

TASK OUTPUTS

A set of Product group/system Profiles to assess:

the market requirements on the manufacturing system, and

the actual performance of the manufacturing system .

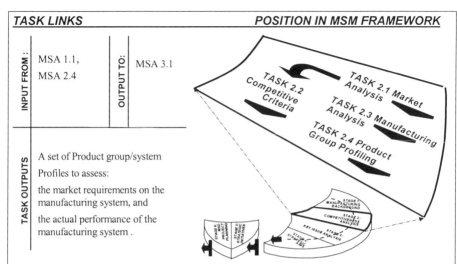

TASK PROCEDURE   TASK FLOWCHART

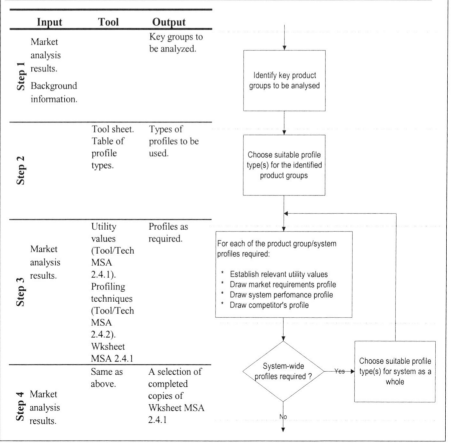

	Input	Tool	Output
Step 1	Market analysis results. Background information.		Key groups to be analyzed.
Step 2		Tool sheet. Table of profile types.	Types of profiles to be used.
Step 3	Market analysis results.	Utility values (Tool/Tech MSA 2.4.1). Profiling techniques (Tool/Tech MSA 2.4.2). Wksheet MSA 2.4.1	Profiles as required.
Step 4	Market analysis results.	Same as above.	A selection of completed copies of Wksheet MSA 2.4.1

Flowchart:

Identify key product groups to be analysed

Choose suitable profile type(s) for the identified product groups

For each of the product group/system profiles required:
* Establish relevant utility values
* Draw market requirements profile
* Draw system perfomance profile
* Draw competitor's profile

System-wide profiles required ? — Yes → Choose suitable profile type(s) for system as a whole

No

TOOLS/TECHNIQUES MSA 2.4.1—*Utility Values*

Suppose in a particular decision situation, a production manager has objectives with respect to the "on time delivery of products" and "average level of WIP" achieved by alternative system layouts. The evaluation process might reveal the following two possible outcomes: outcome of system alternative 1: on time delivery = 85%, average WIP level = 5,000 (items); outcome of system alternative 2: on time delivery = 90%, average WIP level = 8,000 (items). The alternatives tested can be evaluated in terms of different measures of performance because different measures may be regarded as having different utility in comparison with one another. What is required is a means of allowing the performance of the alternatives to be compared across the set of objectives—a method of reducing the multiple outcomes to an overall single measure which will reflect their aggregate utility. For instance, consider a similar decision situation, but now involving two outcomes described in a single unit: outcome (alternative 1): cost of holding WIP = £4,000, cost of machine idleness = £6,000; outcome (alternative 2): cost of holding WIP = £6,400, cost of machine idleness = £5,000. The two system outcomes—the average level of WIP and the average level of machine idleness—are now both described in terms of cost. It can be seen that the system outcome achieved by system alternative 1 is better in terms of WIP, while the outcome from system alternative 2 is better in terms of machine utilization. In spite of this, one would tend to conclude that the outcome associated with alternative 1 is better because it has produced a total cost of £10,000 which is lower than that of £11,400 given by alternative 2. This illustrates the possibility of developing an overall utility indicator if all the outcomes can be assessed on the same ground through arithmetical operations. However, if the outcomes of the same situation is described as previously: outcome 1: on time delivery = 85%, cost of holding WIP = £4,000, cost of idleness = £6,000; outcome 2: on time delivery = 95%, cost of holding WIP = £6,400, cost of idleness = £5,000. Then at least one of the outcomes (the delivery performance) will not be readily suitable for quantitative assessment in terms of cost. When this situation arises, some kind of assessing method must be sought to aid the comparison process. The analysis method of weighted-objectives is one such approach. This analysis procedure involves the following steps:

1) List the project objectives. For example, "objectives **A**, **B**, **C** and **D**".
2) Sort out the objectives in order of importance. Among the objectives listed, some will be considered to be more important than others. This step aims to order the objectives according to how important each is considered to be. A pairs-comparing method may help such a rank-ordering process.

Objectives	A	B	C	D	E	Score
A	0	$s_{a,b}$	$s_{a,c}$	$s_{a,d}$	$s_{a,e}$	S_a
B	$s_{b,a}$	0	$s_{b,c}$	$s_{b,d}$	$s_{b,e}$	S_b
C	$s_{c,a}$	$s_{c,b}$	0	$s_{c,d}$	$s_{c,e}$	S_c
D	$s_{d,a}$	$s_{d,b}$	$s_{d,c}$	0	$s_{d,e}$	S_d
E	$s_{e,a}$	$s_{e,b}$	$s_{e,c}$	$s_{e,d}$	0	S_e

To use this procedure, one picks the first objective from the list, compares it against each of the other objectives, in turn, and records the comparison results $s_{i,j}$ in a chart like the one shown above. The importance indicators, $s_{i,j}$, will be assigned a value of either 0, 1 or 2 depending respectively on whether objective **i** is considered to be less important, equally important or more important than objective **j**. For example, if **A** is considered a more important objective to achieve than **B**, then $s_{a,b}$ should be given a value of 2. If, on the other hand, **A** is considered to be the less important one of the two, then $s_{a,b}$ is 0. This process continues with the second objective in the list, and is repeated until the list is complete. Finally, summing up the scores in a row gives the overall rank order of importance for the associated objective.

*3) **Weight objectives according to rank-order**.* Following the determination of their ranking-order, the next step is to assign a relative weighting factor to each of the objectives. For example, they might simply be allocated along a horizontal axis to show their relative weights:

In this example, objective **C** is shown to be twice as important as **D**, although it is only 70 percent as valued as the most important objective in the list, objective **E** (we will use f_j to indicate such a weighting factor for objective **j**).

*4) **Estimate utility scores for each of the objectives**.* It is next necessary to assign utility scores to the system outcome obtained by the alternatives for each of the objectives. This involves deciphering what a particular level of system outcome means when measured against an objective.

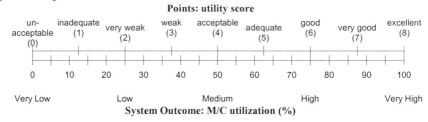

The possible system outcomes are placed on a scale like that shown, and each of the system outcomes is assigned a utility score. This conversion process allows both quantitative and qualitative performance measures to be compared on a similar basis (here we will use $u_{i,j}$ to indicate a utility score for the system outcome given by alternative **i**, against objective **j**).

*5) **Evaluate and compare outcomes using the overall relative utility values**.* Comparison between alternatives can now be made on the basis of their relative utility values. The relative utility value of an outcome given by a particular alternative is obtained simply by multiplying its utility score by its weighting factor. That is, the utility value of the outcome given by alternative **i** against objective **j** is given by:

$$v_{i,j} = u_{i,j} \, f_j$$

As a simple measure of comparison, these individual utility values can be summed up to give a single utility value which indicates the relative overall worthiness of the alternative concerned. Thus the relative worthiness of alternative **i** is given by:

$$V_i = \Sigma \, v_{i,j} = \Sigma \, (u_{i,j} \, f_j)$$

The idea behind this weighted-objective approach—reducing the problem contents to a single dimension—is of great importance in problem solving. It must be emphasized that such a procedure requires skill, experience, and the participation of all parties of the system project in order to succeed.

TOOL/TECHNIQUE MSA 2.4.2—Product/System Profiling

The Relative importance of the Product groups can be established through a set of utility weightings. These are based on a percentage value such that the sum of Relative importance equals one. Against each of the competitive criteria, each Product group is assigned a requirement rating ranging from 0 (not required) to 100

Product group	A	B	C	D
Relative importance (Σ = 1)	0.5	0.	0.13	0.07
Quality	75	80	65	55
Delivery lead-time	50	65	60	15
Delivery reliability	80	70	60	50
Design flexibility	40	90	30	75
Volume flexibility	20	15	80	10
Cost/price	80	25	70	40

(absolutely essential). Hence, if quality is considered to be important, the users may quantify the degree of importance by assigning a value of , say, 75 or 80. The completion of this process allows a profile of the Product groups to be specified, as illustrated in the table. A series of other profiles can be generated from the data to provide additional comparisons within and

Parameter	A	P_i	Ω	Ω_Σ	$\Omega_{\Sigma I}$
Quality	75	37.5	0.22	0.06	0.11
Delivery lead-time	50	25	0.14	0.04	0.07
Delivery reliability	80	40	0.23	0.06	0.12
Design flexibility	40	20	0.12	0.03	0.06
Volume flexibility	20	10	0.06	0.02	0.03
Cost	80	40	0.23	0.06	0.12

between Product groups. For each Product group and competitive criteria pair, the following additional parameters can be calculated:
Relative importance criteria (P_i): criteria value for Product group based on importance of Product group.

Product group normalized criteria (Ω): criteria value for Product group based on ratio of absolute criteria value to sum of all values within same Product group.
Absolute system normalized criteria (Ω_Σ): criteria value for Product group based on ratio of absolute criteria value to sum of all values of all Product groups.
Relative system normalized criteria ($\Omega_{\Sigma I}$): criteria value for Product group based on ratio of Relative importance criteria value to sum of all Relative importance criteria values of all Product groups.

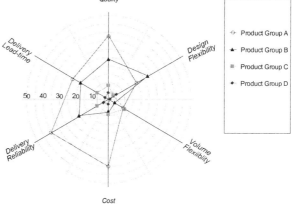

For the example values previously mentioned, the table above gives the values of various profiles for Product group A. Using this technique, it is possible to provide a visual representation indicating the different competitive criteria requirements for each Product group. Each can be described by a vector or represented on a diagram, as illustrated by the relative profile of production group A. The Relative importance criteria values P_i produce the relative Product group profiles, allowing a comparison of Product groups and their criteria, by taking into account their individual contributions to the system. The Product group normalized criteria (Ω) provides an alternative indication of the criteria values of the Product group relative to one another. The absolute system normalized criteria (Ω_Σ) provides an indication of the criteria values of all the Product groups relative to one another. However, the value of this parameter is somewhat limited, given that it does not take into account the Relative importance of each Product group. Conversely, the relative system normalized criteria ($\Omega_{\Sigma I}$) does take into account the Relative importance of each Product group and is therefore a useful alternative parameter for comparisons across Product groups.

In a similar manner, a system profile may be produced, indicating the combined system requirements with respect to the competitive criteria. There are several means by which to establish a system profile. For example, the system profile can be established through the use of an utility function, producing an aggregated utility for each of the criteria based on the Relative importance of the Product groups. The utility profile of the overall system, U, is therefore presented by the vector:

$$U = \{Us,\ Ulf,\ Ur,\ Due,\ Up,\ Us\}$$

where the system's competitiveness value with respect to criteria i is given by:

$$Ui = \sum_{AllGroups} (GroupCompetitivenessValue)_i \times (GroupUtilityValue)$$

For example, according to the above the value of competitiveness with respect to the quality criteria, Us, is given by (assuming only two Product groups A and B):

$$Uq = Qa \times Ia + Qb \times Ib$$

Product group	A	B	C	D	System (U)	
Relative importance	0.5	0.3	0.13	0.07		
Quality	75	80	65	55	Uq	74
Delivery lead-time	50	65	60	15	Ul	53
Delivery reliability	80	70	60	50	Ur	72
Design flexibility	40	90	30	75	Ud	56
Volume flexibility	20	15	80	10	Uv	26
Cost/price	80	25	70	40	Uc	59

...where A is the quality competitive criteria requirement for Product group A, Ian is the Relative importance for Product group A, etc. The results will be a weighted utility profile as shown in the table. Based on the values in the table, the figure below presents the Product group profiles together with the aggregated system utility profile.

Alternative system requirement profiles include:

Maximum Criteria Requirements: The requirements for the aggregated system adopt the maximum requirements from all the Product groups.

Maximum Relative Criteria Requirements: The requirements for the aggregated system adopt the requirements of the Product group with the maximum relative criteria.

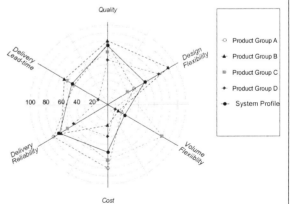

The system performance profiling is similar in detail to that of the competitive criteria stage except that it provides an assessment of how the manufacturing system is actually performing for each Product group with respect to the competitive criteria, rather than the requirements for the system.

WORKSHEET MSA 2.4.1—Product group/System Profiling

Project Title:

Person(s) Responsible:

Version: **Date Completed:**

Type of profile: Product group/system
Type of variables: absolute/relative /absolute-normalized/relative-normalized
Type of requirements: relative-requirement/maximum-relative-requirement

Product group							System
Relative importance							
Quality							
Delivery lead-time							
Delivery reliability							
Design flexibility							
Volume flexibility							
Cost/price							
Other							

2.4 STAGE MSA 3—KEY ISSUES

Having identified the basis for competitive requirements, this stage identifies the key issues that need to be addressed. The successful completion of this should provide an answer to the key question: *where should we be?* The combination of the first three MSA stages can be referred to as *problem formulation* because by establishing *where we are now* and *where we should be*, these stages together will indicate the gap between the present system state and what its environment demands from the system—or a *problem* which prompts the search for an appropriate solution so that the gap can be closed.

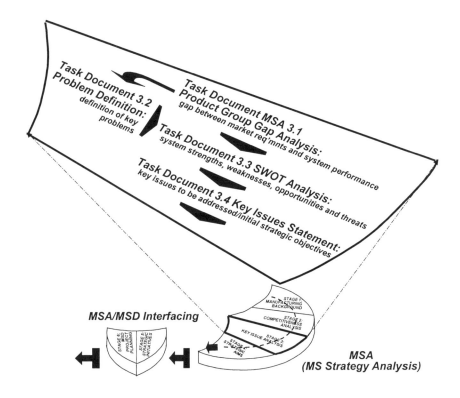

Figure 2.4 Stage MSA 3—key issues

The Product group gap analysis will provide both qualitative and quantitative indications of the differences between what the market and customers require, and the actual performance of the manufacturing system (Figure 2.4). With this information, two options can be followed: either to continue with the strategy capture approach and complete a SWOT analysis with which to derive the key issues, and/or to adopt a problem solving approach and examine a *quick hit* strategy problem chart. The chart itself can be used in conjunction with the SWOT analysis in order to identify key areas for improvement. Following these, key issues for the MS function can be clearly specified. The results of this stage for the example company are summarized below.

MSA 3.1—Requirement/Performance Gap Analysis

The gap analysis of requirements and performance produces the results shown in Table 2.5.

Table 2.5 Example—summary of gap analysis (*Worksheet MSA 3.1.1*)

Gap analysis	Group A	Group B	Group C	Group D	Service A	Service B
Quality	-10	-	20	10	-	5
Delivery lead-time	-10	-45	-35	-20	10	5
Delivery reliability	-	-40	-25	-30	25	10
Design flexibility	-	10	10	-10	10	-
Volume flexibility	-	-25	-15	-20	5	-5
Cost	-30	-20	10	5	5	10

These are also illustrated in Figure 2.5. Similarly, the weighted gap results, using the Relative importance factor, can be calculated as presented in Table 2.6.

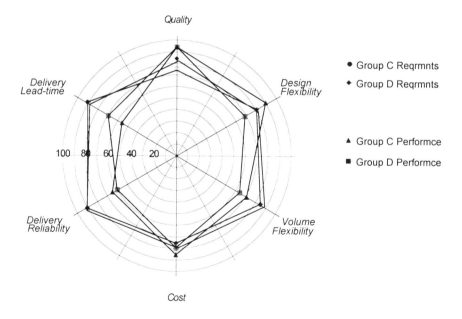

Figure 2.5 Example—Product group requirements/performance profiles

Both the simple analysis and the weighted analysis indicate Delivery lead-time, Delivery reliability and Volume flexibility as the initial targets. This is particularly true considering that, while lead-time is specified as an Order winning criteria for the majority of the Product groups, the system is obviously under-performing in this regard.

Table 2.6 Example—summary of weighted gap analysis (*Worksheet MSA 3.1.1*)

Weighted gap analysis	Group A	Group B	Group C	Group D	Service A	Service B
Importance	20	12	30	35	1	2
Quality	-2	-	6	3.5	-	0.1
Delivery lead-time	-2	-5.4	-10.5	-7	0.1	0.1
Delivery reliability	-	-4.8	-7.5	-10.5	0.25	0.2
Design flexibility	-	1.2	3	-3.5	0.1	-
Volume flexibility	-	-3	-4.5	-7	0.05	-0.1
Cost	-6	-2.4	3	1.75	0.05	0.2

MSA 3.2—Problem Definition
From the quick-hit problem table provided (*Tool/Technique MSA 3.2.1*), the underperformance of lead-time, Delivery reliability and Volume flexibility suggests the main possible problem areas relate to:

under capacity, bottlenecks, and lack of flexibility
lack of coordination, supplier unreliability
inappropriate levels of decision making, ineffective material control
incorrect inventory information,
inappropriate new product introduction process

From the above, the company's own knowledge of the manufacturing system may help it to narrow down the problems (to be recorded in Worksheet MSA 3.2.1) as:

capacity shortage and/or rigid capacity
complex material flow within factory and/or long setup times
inaccurate forecasting and/or incorrect inventory information
subcontractor quality and/or capabilities mismatch

MSA 3.3—SWOT Analysis
The SWOT analysis, with *Worksheet MSA 3.3.1*, gives the results shown in Table 2.7.

MSA 3.4—Key Issues
The key issues for the company, as summarized in *Worksheet MSA 3.4.1* include inadequate forecasting of demand and inadequate capacity, resulting in long lead-times. Using the problem analysis and the SWOT analysis of the previous steps, we may derive the first stage strategic objectives of the company:

to improve forecasting and to improve inventory information
to increase capacity and to increase the workforce skills base
to simplify material flow to reduce setup times
to reassess subcontracting and supplier policies

Table 2.7 Example—summary of SWOT analysis

Threats	Feature	Reason
Economic	Interest rates	hold substantial inventory and raw materials
Social & Political	Government legislation	customs procedures slow company operations
	Environmental legislation	substantial use of water within the processes
Market & Competition	Customer dependence	primarily dependent on a single customer
	Supplier dependence	have one principal steel supplier
Products & Technology	Substitute products	competitors developing a submersible pump
Others	Raw materials	no national natural resources of iron or steel

Opportunities	Feature	Reason
Economic	Availability of credit	government assistance, low interest loans
	Level of employment	easy to recruit and to retain workforce
Demographic	Income levels	everyone receives low pay
	Age composition	relatively smooth between ages of 18 and 60
Market & Competition	Customer plans	customers planning to expand
	Competitor plans	some competitors planning to leave the market
	Supplier plans	suppliers are increasing customer intimacy
Products & Technology	New technology	long life pump with less corrosion
	Substitute products	own design rather than bought in

Weakness	Feature	Reason
Management & Organization	Personnel policies	old system still in operation
Operations	Lead-times	long lead-time products, mainly due to raw materials
	Capacity	employee and machine under capacity
	Volume flexibility	low labor Volume flexibility
	Location	far from export outlets
	Material availability	difficult to obtain raw materials
	Performance	supplier relations and ordering of raw materials need improvement

Strengths	Feature	Reason
Management & Organization	Management systems	good control, computerized facilities, management aims to operate strategically, implementing business process reengineering
	Industrial relations	good relations with the workforce
	Employee age	good range between 19 and 60, mean age of 30
Operations	Quality	adopted ISO 9000 and quality procedures
	Design flexibility	have competent technical engineers
	Dependability	company operates reliably
	Technology	company possesses better technology than national competitors
	Equipment age	company possesses relatively new machines
Finance	Capital structure	some machines have already depreciated,
	Financial planning	finances take into account future conditions
	Accounting system	organized and computerized system
Others	Image of firm	the company has a good reputation for quality

Task Document MSA 3.1—Gap Analysis

TASK OVERVIEW

DESCRIPTION

Gap analysis provides a comparison of the requirements and the actual performance of the MS system for each of the Product groups and, if needed, for the system as a whole. Both tabular and graphical representations can be used for this purpose. In addition to providing a visual representation of a requirement/performance gap, analyses can be carried out to give an indication of the importance of any particular gap. For example, an approximate indication of the improvement required for each Product group with respect to a particular criteria can be established by: $\Delta = N—\Theta$, where: Δ is the gap, N is the required value, and Θ is performance. Hence if $\Delta > 0$ then the system is under-performing for the Product group for a certain criteria. The Relative importance of the gap is given by Δ x Ω, where: Ω is the normalized value as discussed in *Task Document MSA 2.4*. For evaluating the Relative importance of the gap with respect to an individual Product group π, it is suggested that the normalized value Ω_π be applied. For evaluating the Relative importance of the gap with respect to the entire system Σ, the normalized value Ω_Σ should be applied. This normalization procedure provides an indication of the degree of importance of the gaps within each Product group and across the Product groups.

TASK LINKS   POSITION IN MSM FRAMEWORK

INPUT FROM ::	MSA 2.2 MSA 2.4	OUTPUT TO:	MSA 3.2
OUTPUTS		Requirement/performance gap values.	

TASK PROCEDURE   TASK FLOWCHART

Step 1

Input	Tool	Output
Requirement and performance data	Wksheet MSA 3.1.1	A collection of completed Wksheet MSA 3.1.1

For each of the previous product group/system profiles, draw both the requirement and performance data on the same diagram to highlight the possible gaps. Then calculate the requirement/performance gap value(s) against : *quality, delivery lead-time, delivery reliability, product features, design flexibility, volume flexibility, and cost/price*

All relevant profiles considered?   No

Yes

WORKSHEET MSA 3.1.1—Gap Analysis

Project Title:

Person(s) Responsible:

Version: **Date Completed:**

Type of profile: product-group/system
Type of variables: absolute/relative/absolute-normalized/relative-normalized
Type of requirements: relative-requirement/maximum-relative-requirement

For Product group Profiles:
Product group Name/Number: *Product group Importance*:

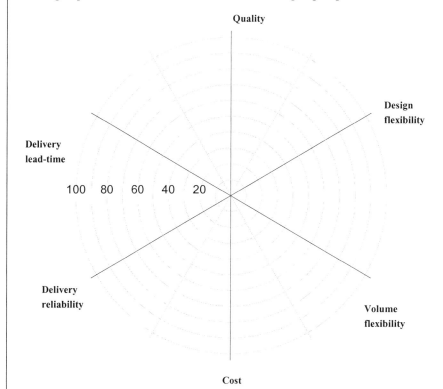

	Absolute Gap Value	Weighted Gap Value—Type:
Quality		
Delivery lead-time		
Delivery reliability		
Design flexibility		
Volume flexibility		
Cost		

Task Document MSA 3.2—Problem Definition

TASK OVERVIEW

TASK DESCRIPTION

Once the performance gaps have been identified, possible causes and reasons for the gaps can be investigated and their relevance to the organization discussed. From the gap analysis results, possible weaknesses and strengths with respect to the six key competitive criteria should be identified. To help this task, a cause-effect table is provided that relates possible problems within the eleven manufacturing policy decision areas. It also outlines their likely effect on the competitiveness of the manufacturing function with respect to the competitive criteria. However, the numeric values and the indications resulting from the gap analysis and the table provide only guidelines. They are intended as a means of stimulating the company management to apply its own intuition and experience. Thus, the tacit knowledge held by the members of the organization concerning the operations of the manufacturing systems also needs to be applied to identify problems.

TASK LINKS   POSITION IN MSM FRAMEWORK

INPUT FROM : MSA 3.1   **OUTPUT TO :** MSA 3.3

TASK OUTPUTS

Identification of possible problems causing the performance gaps, listed under a number of suggested headings.

TASK PROCEDURE   TASK FLOWCHART

	Input	Tool	Output
Step 1	Gap analysis results	Wksheet MSA 3.2.1 Tool/Tech. MSA 3.2.1 (Problem definition table)	A collection of completed Wksheet MSA 3.2.1

For each of the previous product group/ system gaps:

* *Read down the columns in the cause-effect table for each of the competitive criteria.*
* *Read across to the policy area, and identify the possible problems that may cause the good or bad performance.*
* *Record the results under the suggested headings.*

All relevant requirement/ performance gaps considered? — No

Yes

TOOL/TECHNIQUE MSA 3.2.1 —Problem Definition Table

Policy Area	Problems	Quality	Delivery lead-time	Delivery reliability	Design flexibility	Volume flexibility	Cost
Capacity	under capacity	0	Reduced	Reduced	0	Reduced	0
	over capacity	0	Improved	Improved	0	Improved	Reduced
	rigid capacity	0	0	0	0	Reduced	0
	bottlenecks	0	Reduced	Reduced	0	Reduced	Reduced
Facilities	lack of focus	Reduced (?)	Reduced (?)	Reduced	Improved	Improved (?)	0
	too complex	0	Reduced	Reduced	Improved	Improved	0
	lack of integration	0	0	0	0	0	0
	functional layout	0	Reduced	Reduced	Reduced	0	Reduced (?)
	lack of capability	Reduced	Reduced (?)	Reduced (?)	Reduced	Reduced	0
Processes and Technology	lack of capability	Reduced	Reduced (?)	Reduced (?)	Reduced	Reduced	0
	lack of flexibility	0	0	0	Reduced	Reduced	0
	lack of focus	0	0	0	Improved	Improved	0
	functional layout	0	Reduced	Reduced	Reduced	0	Reduced (?)
	long setup times	Reduced (?)	Reduced	0	Reduced	0	Reduced
	no coordination of technology with operations	0	Reduced (?)	Reduced (?)	Reduced (?)	Reduced (?)	Reduced (?)
	low variety, high volume, low integration capability	0	0	0	Improved	Improved	Reduced
	high variety, low volume, high integration capability	0	0	0	Reduced	Reduced	Reduced
	no competitive advantage	0	0	0	0	0	0
	ageing technology	Reduced	0	Reduced (?)	0	0	Reduced
Vertical Integration	lack of focus	Reduced (?)	Reduced	Reduced	Reduced (?)	0	0
	lack of coordination and management	Reduced (?)	Reduced	Reduced	Reduced	Reduced	Reduced
	low ownership of supply chain	0	Reduced	Reduced	Improved	0	0
	high ownership of supply chain	0	Reduced	Reduced	Reduced	Reduced	0
Supplier	unreliable delivery	0	Reduced	Reduced	0	Reduced	Reduced

Policy Area	problems	Quality	Delivery lead-time	Delivery reliability	Design flexibility	Volume flexibility	Cost
Relations	unreliable quality	Reduced	Reduced	Reduced	0	Reduced	Reduced
	increased material costs	0	0	0	0	0	Reduced
	long lead-times	0	Reduced	Reduced (?)	Reduced	Reduced	Reduced
Human Resources	low workforce skill levels	Reduced	Reduced	Reduced	Reduced	0	Reduced
	low supervisor skill levels	Reduced	0	Reduced	0	0	0
	insufficient motivation	Reduced	Reduced (?)	Reduced	0	0	Reduced (?)
	inappropriate level of decision making	Reduced (?)	Reduced	Reduced	0	0	0
	lack of flexibility	0	0	0	0	Reduced (?)	0
	low labor productivity	0	Reduced	0	0	Reduced (?)	0
	direct labor turnover	Reduced	Reduced (?)	Reduced (?)	Reduced	0	Reduced
	ageing workforce	0	0	0	0	0	0
	direct labor absenteeism	0	Reduced (?)	Reduced	0	Reduced (?)	Reduced
Quality Systems	inappropriate product quality	Reduced	0	0	Reduced (?)	0	Reduced
	inappropriate process quality	Reduced	Reduced (?)	Reduced	0	Reduced (?)	Reduced
	inappropriate operations quality	0	Reduced	Reduced	0	0	Reduced
	scrap and waste material	0	0	0	0	0	Reduced
	too much quality documentation	Reduced (?)	0	0	0	0	Reduced (?)
	quality procedures misunderstood	Reduced	0	Reduced (?)	0	0	Reduced (?)
	lack of worker involvement in quality	Reduced (?)	0	Reduced (?)	0	0	0
	lack of ownership of product	Reduced (?)	0	Reduced (?)	0	0	0
	inspection delays	0	Reduced (?)	Reduced	0	0	0
	ageing process technology	Reduced	Reduced (?)	Reduced	Reduced (?)	0	0
Planning and Control	inappropriate level of decision making	Reduced (?)	Reduced (?)	0	0	0	Reduced (?)
	ineffective material control	Reduced	Reduced	Reduced	0	0	Reduced
	high inventories	0	Improved	Improved	Reduced (?)	Improved	Reduced
	incorrect inventory information	0	Reduced	Reduced	0	0	Reduced (?)
	control system too complex	0	Reduced (?)	Reduced (?)	Reduced (?)	Reduced (?)	Reduced (?)
	high overhead costs	0	0	0	0	0	Reduced
	frequent expediting	Reduced (?)	0	Reduced (?)	0	0	Reduced

Category	Item						
New Products and Scope	lack of focus	Reduced (?)	Reduced	Reduced	Improved	0	0
	product line too broad	0	Reduced (?)	Reduced (?)	0	0	0
	too complex	Reduced (?)	Reduced	Reduced (?)	0	0	0
	introductions too frequent	Reduced (?)	0	Reduced	0	Reduced (?)	Reduced
	engineering changes too frequent	0	Reduced (?)	Reduced	0	Reduced (?)	Reduced
	resistance to change	Reduced	Reduced	Reduced	Reduced	Reduced	Reduced
Performance Measurement	poor communication of goals	Reduced	0	Reduced (?)	Reduced (?)	Reduced (?)	Reduced (?)
	inappropriate measures	Reduced	0	Reduced (?)	Reduced (?)	Reduced (?)	Reduced (?)
Organization & Management	poor vertical communication	Reduced (?)	Reduced (?)	Reduced (?)	Reduced (?)	Reduced (?)	Reduced (?)
	poor horizontal communication	Reduced (?)	Reduced (?)	Reduced (?)	Reduced (?)	0	Reduced (?)
	poor communication of strategy	Reduced (?)	Reduced (?)	Reduced (?)	Reduced (?)	Reduced (?)	Reduced (?)

Key — For each criteria:

Reduced	Ability to be competitive reduced
Reduced (?)	Ability to be competitive possibly reduced
Improved	Ability to be competitive improved
Improved (?)	Ability to be competitive possibly improved
0	Insignificant effect envisaged

WORKSHEET MSA 3.2.1—Problem Definition

Project Title:

Person(s) Responsible:

Version: **Date Completed:**

Type of profile: product-group/system
Product group name/No (for product group based analysis):

	Gap	Reasons
Quality		
Delivery lead-time		
Delivery reliability		
Design flexibility		
Volume flexibility		
Cost		

Note: The cause-effect table of *Tool/Techniques MSA 3.2.1* gives an indication of possible problem areas within the current manufacturing strategy and manufacturing systems. By reading down the columns in the table for each of the competitive criteria, and then reading across to the policy area, possible problems affecting performance may be suggested. For example, a reduction in competitiveness with respect to Delivery lead-times, Delivery reliability and Volume flexibility may be caused by capacity policies that result in the system being *under capacity*. The table above can then be used to record the problems under a number of the suggested headings.

Task Document MSA 3.3—SWOT Analysis

TASK OVERVIEW

DESCRIPTION

Following the issues of systems thinking discussed in Chapter 1, a SWOT (strength, weakness, opportunity and weakness) analysis serves as a means of matching the environmental threats and opportunities with the company's weaknesses and strengths. It is essentially a creative process of qualitative analysis, and refers to both the internal and external environments. The internal analysis serves to pinpoint the strengths and weaknesses of the organization, involving identifying the quantity and quality of resources available to the MS function. The external analysis, on the other hand, identifies strategic opportunities and threats in the organization's operating environments. These environments include both the immediate industrial environment in which the organization operates and the wider macro-environment.

TASK INSTRUCTION

The relevant issues and factors are summarized as follows:

Strengths: activities, processes, technologies, procedures, etc., which the manufacturing organization does uniquely well.

Weaknesses: activities, processes, technologies, procedures, etc., which the organization does not do to an acceptable standard.

Opportunities: activities, processes, technologies, procedures, events, potential events, etc., which the organization may additionally exploit.

Threats: activities, processes, technologies, procedures, etc., which may prevent the organization reaching its goals.

Threats and opportunities relate to the external environment of the manufacturing organization under analysis, while weaknesses and strengths relate to the internal environment. The analysis can be carried out at various levels along the organizational hierarchy, depending on the level of abstraction required and the factors being addressed. However, if the analysis is to be meaningful, generally it should be used in a disaggregated manner, ideally at the Product groups level or, if necessary, at the individual product level.

To a certain extent, the analysis involved can be a structured process. The analysis is achieved principally through the use of sub-headings, under which specific points and details can be written as shown in the table of *Tool/Technique MSA 3.3.1*. Each of the SWOT categories listed in this generic table should be considered in turn using the following steps:

1) Take each of the headings from the table, and decide whether these are relevant in the particular situation.

2) Provide explanation or justification for each SWOT assessment, indicating the nature and extent of each SWOT, and provide detailed data to support the justification.

3) Further identify key issues by requirement/performance comparison.

4) For strengths and weaknesses, define what the strengths should be and what weaknesses the MS function must avoid.

5) For opportunities and threats, define what opportunities the MS function must take advantage of.

However, sub-headings such as those given in this table should not be seen as an exhaustive list and, where applicable, additional items may be appended. Supporting evidence and data should also corroborate each decision.

TASK LINKS

INPUT FROM:	OUTPUT TO:
MSA 3.1 MSA 3.2	MSA 3.4

TASK OUTPUTS

Identification of opportunities and threats (external factors) to the organization, and strengths and weaknesses of the organization (internal factors), listed under a number of suggested headings.

POSITION IN MSM FRAMEWORK

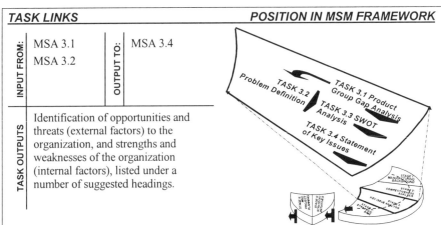

TASK PROCEDURE

	Input	Tool	Output
Step 1	Gap analysis results	Wksheet MSA 3.3.1	
	Problem identification results	Tool/Tech. MSA 3.3.1	
Step 2	Data from previous stages. Additional internal data, as required.		
Step 3	Data from previous stages. Additional external data, as required.		A collection of completed Wksheet MSA 3.3.1.

TASK FLOWCHART

For each of the chosen product groups/system to be considered:

Decide which of the items in the sub-heading table are relevant and, for those which are relevant, identify whether it is a strength, weakness opportunity or threat.

For each of the strengths and weaknesses:

Decide what the company's strength should be, and what weakness must be eliminated.

For each of the opportunities and threats:

Decide what opportunities must the company take advantage of, and what threats need attention.

All chosen product groups/systems considered? — No

Yes

TOOL/TECHNIQUE MSA 3.3.1—SWOT Sub-Heading Table

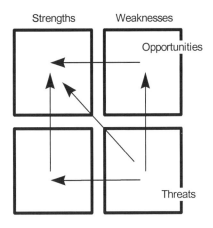

Strengths   Weaknesses

Opportunities

Threats

The overall aim in a SWOT exercise is to identify future strategic directions that will effectively direct the organization in such a way so that the center of attention is as shown in the figure. The table below provides a list of typical sub-headings. The first set of headings relates primarily to the opportunities and threats. The second set relates primarily to the strengths and weaknesses.

Opportunities and threats	Strengths and weaknesses
Economic factors	**Management & organizational factors**
Interest rates	Management systems
Exchange rates	Industrial relations
Availability of credit	Personnel policies
Level of employment	Morale
Social and political factors	Skills
Government legislation	Employee experience
European legislation	**Operations**
International legislation	Quality
Union plans	Lead-times
Consumer groups	Performance
Special interest groups	Capacity
Environmental issues	Flexibility
Demographic factors	Dependability
Demographics	Location
Income levels	Material availability
Age composition	Technology
Market and competition criteria	Equipment age
Customer plans	Implementing change
Competition plans	**Finance factors**
Supplier plans	Capital structure
Customer dependence	Profitability
New competitors	Financial planning
Supplier dependence	Accounting system
Products and technology	Cost structure
New products	**Other factors**
New markets	Patents
New technology	Image of firm
Substitute products	
Other factors	
Availability of raw materials	

WORKSHEET MSA 3.3.1 —SWOT Analysis

Project Title:

Person(s) Responsible:

Version: **Date Completed:**

Type of profile: product-group/system
Product group Name/No (for product group based analysis):

Opportunities/Threats	Possible Action
Economic factors 1) 2) 3)	
Social and political factors 1) 2) 3)	
Demographic factors 1) 2) 3)	
Market and competition criteria 1) 2) 3)	
Products and technology 1) 2) 3)	
Strengths/Weaknesses	Possible Action
Management/organization 1) 2) 3)	
Operations 1) 2) 3)	
Finance factors 1) 2) 3)	
Other factors	

Task Document MSA 3.4—Statement of Key Issues

TASK OVERVIEW

DESCRIPTION

This task records relevant information concerning events, trends and facts which have an impact on the organization. In particular it covers any key issues arising from the problem definition and the SWOT analysis. Their implications and effects on the MS operation are also recorded. Finally, the preliminary strategic objectives are derived from the key issues.

TASK LINKS

POSITION IN MSM FRAMEWORK

INPUT FROM :
Problem definition results.
SWOT analysis results.

OUTPUT TO:
MSA 4.1

OUTPUT
Textual statement of key issues identified from the previous analysis.

WORKSHEET MSA 3.4.1 —Key Issue Statement

Project Title:

Person(s) Responsible:

Version: **Date Completed:**

Key manufacturing issues

Key issues arising from external SWOT analysis:

Key issues arising from internal SWOT analysis:

Key supply chain issues

Regarding suppliers:

Regarding warehouses/depots/transportation:

2.5 STAGE MSA 4—STRATEGIC AIMS

Having identified a gap that needs to be filled, the logic of a journey planning process then requires the answer to two more questions: *what are the possible routes and means?;* and *which route to take?* The rest of the MSA/MSD cycle aims to identify the feasible alternatives and analyze the possible consequences of each of these routes. This allows one to choose the strategy that best satisfies the particular requirements as identified through stages MSA 1 to 3. Therefore, the overall aim of this stage is to transform the problem definitions into strategic aims, from which strategic initiatives and action plans can be derived. Highlighting, in particular, those aspects that the subsequent MSD project(s) must deal with, the aims of this stage may be summarized as follows:

- To assist in defining the problems and root causes of problems related to the operations of the current MS system, and
- To define the starting point from which the future manufacturing strategy will emerge. This provides a means of assisting the evolution of action plans and of indicating the direction in which the MSD project is to develop.

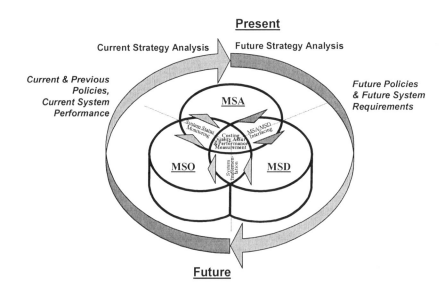

Figure 2.6 Time line of the MSA/MSD cycle

From an application's point of view, the MSA/MSD/MSO cycle of the MSM framework provides a basis for studying the evolution of MS strategies over time, as shown in Figure 2.6. In fact, if the organization under study already has a well developed and documented MS strategy, then this stage may be considered an alternative "entry point" into the MSA/MSD/MSO cycle. Based on this cycle, an analyst will also have the opportunity to return to this stage of the analysis throughout the subsequent stages in order to analyze and assess the implications and the impact of the current strategic decisions. Hence, not only can this stage be the initiation point of an MSD project, but there is also the option of either capturing the present policies and formulating future policies. Consequently, this stage consists of four tasks that are grouped into two parallel sections, as shown in Figure 2.7.

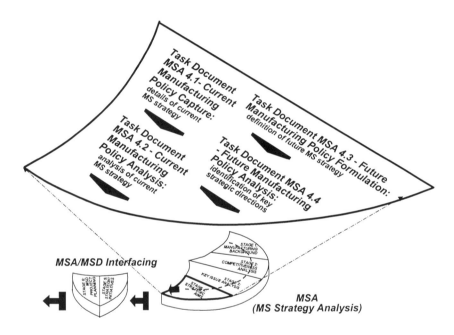

Figure 2.7 Stage MSA 4—strategic aims

Again, the example company will be used to illustrate the procedures involved. Although a current documented manufacturing strategy for this example company is not available, a series of manufacturing policies or practices in use can be extracted. These policies are captured and examined in order to assess how adequately they meet the requirements of the business and the manufacturing function.

MSA 4.1—Current Policy Capture
Table 2.8 summaries the current manufacturing policy of the example company, as captured using *Worksheet MSA 4.1.1*.

Table 2.8 Example—current manufacturing policies/practice

Policy Area	Policies
Capacity	Pitched at average demand, rapid capacity expansion required, minimum economic floor space, plant capacity uses three shifts running for 24 hours, subcontract for demand highs, expansion through new equipment.
Facilities	Separated plants/split sites, cellular manufacturing focused on processes, simplifying material flow, medium manufacturing integration.
Processes & Technology	Flexible machining centers, high capital intensity, batch manufacture.
Vertical Integration	Low ownership integration, suppliers are subcontractors, when capacity meets demand will reduce subcontracted work, Kanban control of suppliers.
Supplier Development	Close links developed with suppliers, strong reliance on suppliers for subcontracting (due to demand increase), development of Kanban control with suppliers, suppliers to be as "local" as possible, still relatively competitive.
Human Resources	Job skills improvement, general purpose teams, recruit qualified staff.
Quality Systems	SPC, quality circles, in-process inspection, ISO 9000.
Planning & Control	Reduce inventory, Kanban control.
Product Scope & New Products	Originally planned to cease production of old product to make way for new product introduction. Demand for both products has increased. QFD, concurrent engineering utilized.
Performance Measures	Business ratios.
Organization	Hierarchical and functional, manufacturing is relatively flat with cell leaders and cell operators

MSA 4.2—Current Policy Analysis

The analysis of the current policies indicates that (*Worksheet MSA 4.2.1*):

- Capacity policies suggest a negative effect on Delivery lead-times, reliability and Volume flexibility.
- Facilities policies suggest a slight negative effect on Delivery lead-times and reliability.
- Process and technology are seen as having a restriction on volume and Design flexibility.
- Vertical integration is seen as having very little effect.
- Supplier development policies suggest a negative effect on quality and lead-times.
- Human resources are seen as having a negative effect on quality due to the lack of skills.
- Quality systems are seen as having a slight positive effect on quality and Delivery reliability.
- Production planning and control suggest a positive effect on costs, but little effect elsewhere.
- Product scope and new products policies suggest a slight negative effect on Delivery lead-time and Delivery reliability, but a slightly positive effect on Design flexibility.
- Performance measures are seen as potentially having a positive effect.
- Organization policies are seen as having positive effects on costs and quality.

MSA 4.3—Future Strategy Formulation

Table 2.9 indicates the key policy changes captured for the future strategy.

Table 2.9 Example—future manufacturing policies

Policy Area	Policies
Capacity	Increase capacity through new equipment and new facility.
Facilities	New site development, adopt cellular manufacture where it is beneficial, trying to simplify material flow, split site between core businesses.
Processes & Technology	Single hit manufacture, apply technology only for the benefits, adopt standard modular machine tools rather than expensive flexible machine tools.
Vertical Integration	Not an issue.
Supplier Development	Change policy to farm out volume bits to subcontractors and not difficult bits.
Human Resources	Develop job skills, increase quality concern, general purpose teams, recruit qualified staff.
Quality Systems	Quality program, SPC, quality circles, in-process inspection, ISO 9000.
Planning & Control	Reduce inventory, improve control, simplify material flow, improved capacity planning required.
Scope and New Products	QFD, concurrent engineering.
Performance Measures	Business ratios.
Organization	No major change in human organization.

MSA 4.4—Future Policy Analysis

This stage involves the application of the strategy relationship tables (*Tool/Technique MSA 4.1.2*) to ensure that the policies captured are consistent and coherent. For example, when considering the capacity policy with respect to expanding capacity, the company should consider also how it relates to the decisions made concerning facilities location, specification and functional integration, type of equipment and process focus, vertical integration and labor policies. Additionally, the policies are assessed with respect to their degree of compliance with the strategic objectives, key issues and problems identified and their contribution to the competitiveness of the manufacturing function. The results of the analysis of the future policies using *Worksheet MSA 4.2.1* indicate that:

- Capacity policies may have positive effects on design and Volume flexibility, but a slight negative effect on costs.
- Facilities policies have positive effects on Delivery lead-times and reliability.
- Process and technology policies may have positive effects on Delivery lead-times and reliability, and design and Volume flexibility.
- Vertical integration policies may have very little effect.
- Supplier development policies were seen as potentially having a slightly positive effect on quality, Delivery reliability and costs.
- Human resources policies may have a slightly positive effect on quality.
- Quality systems policies have a positive effect on quality, Delivery reliability.
- Production planning has a positive effect on costs and Delivery reliability.
- Product scope and new products policies may have very little effect.
- Performance measures were seen as potentially having a positive effect.
- Organization policies may have slight positive effects on costs and quality.

Task Document MSA 4.1—Current Policy/Decision Capture

TASK OVERVIEW

DESCRIPTION

This task aims to help identify the current key strategic polices or decisions from a list of eleven generic areas. These decisions may be those contributing towards the strengths or weaknesses of the MS function, or general decisions that the company recognizes as being potentially strategic in nature. In order to assist in this process, a series of questions and possible response options are provided. This questionnaire approach aims to capture the detailed contents of the current manufacturing strategy. Further help is provided through a strategy-relationship table that suggests possible influences and relationships between sub-decisions across the eleven strategic policy areas. Together, these provide a coherent approach to identifying and capturing the contents of strategic decision-making. The results from this task will provide a detailed account of the company's current practice in the key areas of concern.

TASK INSTRUCTION

Based on the results from previous stages, the analysis commences with the identification of the key decisions within the eleven policy areas, producing an assessment of the operations and infrastructure of the current MS system. The process depends on capturing relevant information covering the details of the MS policies, and hence relies on the application of the questionnaire to structure the approach.

Identify key decisions and the related current practice of the company. A database of decisions and options are provided. For each policy area, there are a number of decisions associated with it, and each of these have several questions. The users may select pre-defined responses to these questions or enter their own responses. In addition, they may enter information concerning a policy area that has not been included in the decisions and options database mentioned previously. The MS strategy and associated policy areas are likely to address the MS system as a single entity, particularly when the current manufacturing strategy and existing operating policies are concerned. As such, it is appropriate to capture the MS strategy and policy information with respect to the entire manufacturing function. However, the approach encourages and offers the opportunity for users to input strategic information for individual Product groups, if such information is available. Since responding to over 200 policy questions can be a demanding exercise, the user is encouraged to prioritize and focus his/her approach on the more important policy areas. The user is able to refer back to the outputs from the previous stage, in particular the quick-hit table and the actions derived from the SWOT analysis. These will provide an indication of which policy areas could be addressed first as a means of maximizing the effect of the effort put into this stage. A priority table is included in the worksheet to assist this process.

Identify interrelated issues to the above, and the related current company practice. Once the set of pre-defined questions is answered, the analyst can refer to the decisions relationship table to identify the possible influences between strategic decisions. For each sub-decision or question, the table indicates possible decision headings throughout all eleven policy areas. The influences could additionally be considered in parallel due to the existence of possible interrelationships. For example, a response to the question concerning the specification of total plant capacity would suggest the strategy also consider the manufacturing policy decisions related to total capacity, demand forecasting, the number of facilities, the specification of facilities, process organization, the position in the supply chain, and make versus buy issues. The table provides generic and intuitive information derived from the literature and, as such, provides guidance to assist the thinking process and creativity involved in strategy capture and formulation. Not only can it be used as a guide to answering policy questions, it can also be used to refer back to questions previously answered which may have a bearing on the decision currently under consideration.

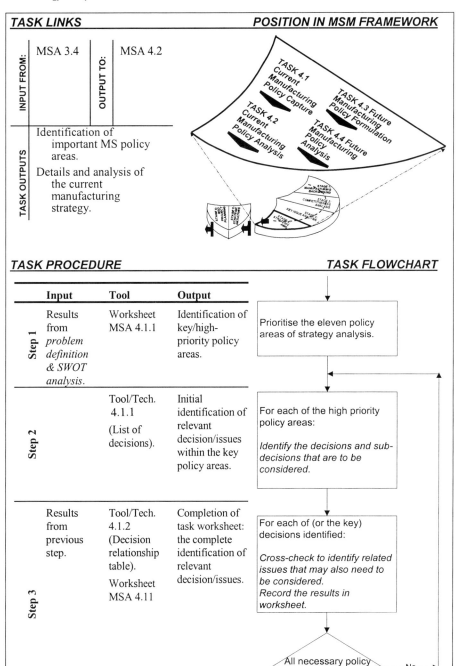

TASK LINKS

INPUT FROM: MSA 3.4

OUTPUT TO: MSA 4.2

TASK OUTPUTS

Identification of important MS policy areas.

Details and analysis of the current manufacturing strategy.

POSITION IN MSM FRAMEWORK

TASK 4.1 Current Manufacturing Policy Capture

TASK 4.3 Future Manufacturing Policy Formulation

TASK 4.2 Current Manufacturing Policy Analysis

TASK 4.4 Future Manufacturing Policy Analysis

TASK PROCEDURE

	Input	Tool	Output
Step 1	Results from *problem definition & SWOT analysis*.	Worksheet MSA 4.1.1	Identification of key/high-priority policy areas.
Step 2		Tool/Tech. 4.1.1 (List of decisions).	Initial identification of relevant decision/issues within the key policy areas.
Step 3	Results from previous step.	Tool/Tech. 4.1.2 (Decision relationship table). Worksheet MSA 4.11	Completion of task worksheet: the complete identification of relevant decision/issues.

TASK FLOWCHART

Prioritise the eleven policy areas of strategy analysis.

For each of the high priority policy areas:

Identify the decisions and sub-decisions that are to be considered.

For each of (or the key) decisions identified:

Cross-check to identify related issues that may also need to be considered.
Record the results in worksheet.

All necessary policy areas analysed? — No

Yes

TOOL/TECHNIQUE 4.1.1—Decisions and Options

CAPACITY

Demand Pitch

How has the total manufacturing capacity been pitched relative to demand?

How have the individual manufacturing capacities been pitched relative to demand?

How has the total capacity been specified with respect to floor space?

How has the total capacity been specified with respect to plant?

How has the total capacity been specified with respect to equipment?

How has the total capacity been specified with respect to labor?

Variation Satisfaction

How have cyclical demand variations been managed?

How have long-term demand variations been managed?

How have demand highs been satisfied?

How have demand lows been satisfied?

What was the degree of flexibility in capacity envisaged for manufacturing?

Expansion Methods

What methods have been used for expanding capacity?

What has been the size of expansion increments?

What has been the trigger for the decision to expand capacity?

Contraction Methods

What methods have been used for contracting capacity?

What has been the size of contraction decrements?

What has been the trigger for the decision to contract capacity?

Timing

How has the timing of capacity changes been determined with respect to demand?

Bottlenecks

Are there any significant bottlenecks that have been identified?

Demand Forecasting

How has demand been monitored?

How has demand been forecasted?

What have been the capacity change signals?

Implications

What have been the implications of capacity for manufacturing?

FACILITIES

Specification

How many facilities have there been?

How has the size of each facility been determined?

What has been the capability of each facility?

Location

What has primarily determined the location of the factory?

What has primarily determined the location of the individual production facilities?

What has primarily determined the location of the central/regional/main distribution centers?

What has primarily determined the location of the individual warehouses?

What type of plant layout has been adopted?

What type of warehouse layout has been adopted for each inventory holding facility?

Focus

What has been the degree of specialization of the facilities?

What has determined the type of focus or specialization of the facilities?

What has been the degree of flexibility of the facilities?

Function Integration

What has been the degree of functional integration within the enterprise?

What has been the degree of functional integration within the manufacturing function?

What has been the degree of functional integration with the logistics services?

Flow

What degree of emphasis has been placed on the flow of materials within each facility?

What degree of emphasis has been placed on the flow of information within each facility?

Implications

What have been the implications of facilities for manufacturing?

PROCESSES AND TECHNOLOGY

Type of Equipment

What has been the degree of flexibility of the production/material-handling/transportation equipment?

What has been the degree of capital intensity of the production/material-handling/transportation equipment?

What has been the degree of capability of the production/material-handling/transportation equipment?

What has been the degree of mechanization of the production/material-handling/transportation equipment?

What has been the degree of automation of the production/material-handling/transportation equipment?

What has been the degree of integration of the production/material-handling/transportation equipment?

What has been the policy with respect to key technologies?

What degree of technological risk has been adopted?

What has been the degree of process innovation adopted?

How have setups and changeovers been satisfied?

What has been the degree of labor intensity of the production/material-handling/transportation equipment?

What has been the degree of maintenance required for the production/material-handling/transportation equipment?

What has been the degree of supervision required for the production/material-handling/transportation equipment?

Process Organization

What type of manufacturing process choice has been adopted?

Focus

What degree of specificity has been adopted?

Man-machine Interface

What has been the extent of job content between machines and manpower?

What has been the extent of skills required by the workforce?

Implications

What have been the implications of processes and technologies for supply chain?

VERTICAL INTEGRATION

Supply Chain Ownership

What has been the degree of ownership of the supplier network?

What has been the degree of ownership of the customer network?

What has been the type of ownership of the supply chain?

What has been the degree of management of the supply chain?

What has been the degree of coordination of the supply chain?

What transaction mechanisms have been adopted for the supply chain?

Expansion and Contraction

What has been the primary means of expanding the supply chain?

What has been the primary means of contracting the supply chain?

Position in Chain

What has been the degree of focus with respect to the position in the supply chain?

How have vertical integration decisions affected supplier relations?

How have vertical integration decisions affected distributor relations?

How have vertical integration decisions affected customer relations?

Implications

What have been the implications for make versus buy decisions?

What have been the implications of vertical integration for the manufacturing function?

SUPPLIER RELATIONS

Competitive Type

What type of relationship has the organizational supply chain logistics function had with its suppliers?

Time Span

What has been the time span of supplier relationships?

Sourcing

What sourcing policies have been adopted?

Supplier Qualification

What means of supplier qualification have been adopted?

How has the performance of suppliers been measured?

How have suppliers been controlled?

What selection criteria have been used for suppliers?

Partnerships

What types of supplier partnerships have been adopted?

What degree of assistance has been given to suppliers?

What degree of technological cooperation has been given to suppliers?

What degree of integration has there been with the suppliers?

What type of integration has there been with the suppliers?

What type of communication has there been with suppliers?

Make versus Buy

What components have been bought?

What services have been bought?

Implications

What are the implications of supplier relations for the organization's supply chain function?

HUMAN RESOURCES

Cultural Properties

What type of human behavior has been encouraged within the logistics/manufacturing function?

What degree of supervision has been suitable?

What type of interdependence has been suitable?

What degree of risk taking has been encouraged?

What has been the degree of ownership of the processes?

What has been the degree of ownership of the products?

What degree of responsibility has been encouraged?

What has been the degree of comfort within the organization?

What type of teams has been formulated?

What has been the extent of communication within the organization?

Production Related

What has been the degree of concern for quality?

What have been the means of controlling quality?

What has been the degree of concern over the processes?

What has been the degree of concern for productivity?

What has been the degree of flexibility and change of the workforce?

What has been the degree of job content?

What has been the extent of the cycle times?

What have been the means of pacing the work?

What has been the level of skills required?

What have been the methods of training adopted?

How have employees been motivated?

General

What has been the degree of employment security?

What has been the policy with respect to overtime?

What has been the policy with respect to employee selection?

What has been the policy with respect to employee recruitment?

How many shifts have been maintained?

What has been the policy with respect to safety issues?

What has been the policy with respect to health issues?

Remuneration

What payment systems have been adopted?

What payment structures have been adopted?

What has been the range of payments available?

What incentives and rewards schemes have been adopted?

Implications

What have been the implications of human resource policies for the supply chain logistics and manufacturing function?

QUALITY SYSTEMS

Implementation

What has been the extent of quality systems implementation?

Process Quality

What has been the degree of capability versus inspection?

What means have been adopted to implement capability and/or inspection?

What have been the functions of inspection processes?

What has been the frequency of inspection?

What quality training has been provided?

How has quality been monitored?
Total Quality
What total quality initiatives have been adopted?
What level of documentation has been adopted?
What aspects of total quality training have been adopted?
Where has the responsibility for total quality been within the organization supply chain?
Quality Levels
How have quality levels been selected?
What have the quality levels been?
Implications
What have been the implications of quality policies for the supply chain logistics and manufacturing function ?

PRODUCTION PLANNING AND CONTROL
Supplier Relations & Inventory
What has been the inventory policy with respect to the suppliers?
What has been the degree of inventory holdings?
What has been the degree of spread of inventory?
What has been the degree of balance?
Where has inventory been located?
What has been the function of inventory?
Manufacturing Priorities
What methods have been adopted to determine manufacturing priorities?
What level within the organization have manufacturing priorities been determined?
What has been the degree decentralization with respect to manufacturing priorities?
What has been the degree coordination with respect to manufacturing priorities?
What has been the degree autonomy of with respect to manufacturing priorities?
What has been the degree of response with respect to manufacturing priorities?
Management
What methods and philosophies have been adopted for materials management?
What has been the attitude with respect to customer promises?
What has been the attitude with respect to customer order changes?
Forecasting
What systems have been adopted for forecasting of demand?

What has been the level of investment in forecasting demand?
Planning
What has been the time horizon adopted for production planning?
What has been the degree of formality of productions planning?
Scheduling
What has been the time horizon adopted for production scheduling?
What have been the policies for resource allocation?
What formal scheduling paradigms have been adopted?
What informal methods of scheduling have been permitted?
What has been the degree of centralization with respect to scheduling?
What has been the degree of monitoring of production?
What has been the scheduling timeframe updating period?
Control
What control policies have been adopted?
What policies have been adopted for the release of orders?
What policies have been adopted for expediting?
What policies have been adopted for batch sizes?
Implications
What has been the approach adopted for production with respect to supply chain structure?
What have been the implications of production planning and control for manufacturing?

PRODUCT SCOPE AND NEW PRODUCT INTRODUCTION
Product Details
What has been the degree of scope of products manufactured?
What has been the degree of focus of products manufactured?
What has been the range of products manufactured?
What has been the volume of products manufactured?
Introduction
What has been the rate of new product introductions?

What philosophies have been adopted for the introduction of products?

What has been the typical life cycle duration of products?

What computer aids has been adopted to assist product introduction?

What has been the extent of computer assistance?

What degree of innovation has been adopted within the organization?

Lead-times

What has been the extent of product design lead-times?

What has been the extent of manufacturing lead-times for new products?

Implications

What have been the implications of product scope and new products for supply chain logistics and manufacturing?

PERFORMANCE MEASUREMENT

General

What selection criteria have been adopted for performance measurement?

What has been the degree of focus on competitive variables?

What has been the degree of focus on business management integration?

What has been the attitude towards benchmarking?

What has been the extent to which performance measures drive strategy?

How explicit have the logistics/manufacturing performance measures been?

How formal have the logistics/manufacturing process measures been?

How formal have the supply chain/manufacturing output measures been?

What has been the extent of feedback of performance measures to supply chain management?

What has been the extent of feedback of performance measures to supply chain operators?

To what extent have performance measures been aimed at the development of capabilities?

What has been the balance between financial and non-financial performance measures?

What has been the reliance on internal measures of performance?

What has been the reliance on external measures of performance?

What type of data has been recorded?

Where has the data been measured within the organization?

Implications

What has been the implication of performance measurement with respect to supply chain logistics and manufacturing?

ORGANIZATION

Structure & Management

What has been the overall structure of the organization?

What has been the degree of openness of management?

What has been the degree of product understanding of management?

What has been the degree of manufacturing understanding of management?

What has been the degree of systems perspective adopted by management?

What has been the culture adopted by management?

Functions

Where has the functional emphasis laid within the manufacturing organization?

What has been the degree of management supervision adopted?

Coordination

What has been the degree of coordination with marketing?

What has been the degree of coordination with engineering?

What has been the degree of coordination with the customers?

Implications

What have been the implications of organization with respect to the manufacturing function?

TOOL/TECHNIQUE 4.1.2—Decisions Relationship Table

Decisions / Sub-decisions	CAPACITY · Total capacity	Variation Satisfaction	Expansion Methods	Contraction Methods	Timing	Bottlenecks	Demand forecasting	FACILITIES · Number	Specification	Location	Focus / specialisation	Function Integration	PROCESSES · Material Flow	Information flow	Type of Equipment	Degree of competitive appl	Material handling	Process Organisation	Focus	Man-machine interface	VERTICAL · Supply chain ownership	Expansion	Contraction	Position in chain	Make vs buy implications	SUPPLIER · Competitive type	Time span	Sourcing type	Supplier	Products	HUMAN RE · Cultural properties and cor	Production related	General	Remuneration	Implementation	QUALITY · Product quality	Process quality	Total quality	Quality levels	PROD PLANNING · Supplier relations	Inventory	Manufacturing priorities	Management	Forecasting	Planning	Scheduling	NEW PRD · Production approach	Product details	Introduction	Lead-times	Performance	ORGNSA'N · Structure	State	Management	Functions	Co-ordination
Total capacity																																																								
C Demand pitch							X	X	X															X	X																			X												
A Floor Space							X	X	X															X	X																															
P Plant							X	X	X						X									X	X																															
A Equipment							X																	X	X																															
C Labour							X																	X	X																															
I Variation																																																							
T Cyclical							X																										X								X			X	X										
Y Long Term Trends																																	X								X				X	X									
Demand Highs							X																						X				X								X	X	X	X	X	X									
Demand Lows																													X				X								X				X	X									
Degree of flexibility																																									X				X	X									
Expansion																																																								
How																			X			X											X																							
Size of increment		X																				X	X										X																						
Contraction																																																								
How					X																		X										X									X													
Size of decrement																							X																				X												
Timing	X																																																						
Bottlenecks	X					X																																			X														
Forecasting																																																							
How monitor	X						X																																					X	X					X	X				
How forecast							X																																					X	X					X	X				
Cap. change signal		X		X																																								X	X			X		X	X			X	
Implications																																																								
investment	X	X	X	X				X	X						X			X																								X			X									X
cost and payback	X	X	X					X	X						X																												X		X									
service levels	X	X	X	X																																							X		X	X									
risk				X																											X																				X	X	X	X		
workforce mngmnt																																																			X			X	X	
quality				X																																		X	X												X			X	X
fulfilling manuf role				X																																			X												X	X			X

(Continued on the following pages)

	Category	Attribute
F		Number
A	**Specification**	
C		Size
		Capability
L	**Location**	
T		Factory
		Facilities
I		Plant Layout
E	**Specialisation**	
S		Type
		Degree
	Integration	
		Enterprise
		Manufacturing
		Support services
	Material Flow	
		Informatn flow
	Implications	
		Costs
		Risks
		Service levels
		Organisation
		Control
		Workforce mngmt
		Schd'lg/inventory
		Fulfiling mant'g role
	Type of Equip't	
P		Flexibility
R		Capital intensity
O		Capability
C		Mechanisation
E		Automation
S		Integration
S		Key Technologies
E		Technological risk
S		Rate of process innov
		Setup requirmt
		Mechnisation
		Maintenance required
		Supervision required
		Balance of capacities
		Equipment cost
	Competitiveness	
		Tooling
		Equipment
		Manftg engineering
		Industrial engineering
	Material handling	
		Automation
		Integration
	Process Org'n	
		Process-volume
		Focus
	Man-M/C interface	
		Job content
		Skills required

	Cultural&control																																																							
H	Behaviour									x	x	x			x			x	x	x													x	x		x	x	x	x	x			x				x							x	x	x	
U	Time horizon									x		x			x			x	x	x													x										x				x							x	x	x	
M	Supervision									x	x	x						x		x													x						x				x	x								x	x	x	x	x	
A	Interdependence									x	x	x		x		x		x		x													x			x	x	x	x	x			x	x			x					x	x	x	x	x	
N	Risk taking attitude								x	x					x	x				x													x	x	x	x			x				x				x					x		x			
	Ownership of process									x					x	x		x											x			x			x	x	x	x				x						x			x		x	x	x	
R	Ownership of product									x						x		x											x			x			x	x		x				x						x	x		x		x	x	x	
E	Responsibility									x	x	x				x															x	x	x	x	x	x	x	x			x	x		x	x						x	x	x	x	
S	Comfort		x							x	x	x	x		x	x	x		x	x	x											x	x									x	x			x			x			x	x	x	x	
O	Teams									x	x	x		x	x	x		x		x													x	x	x	x			x				x				x					x	x	x	x	x	
U	Communication													x						x													x										x										x	x	x	x	x	x
R	**Production related**																																																							
C	Quality concern																					x											x				x	x	x	x	x			x										x	x	x		x	
E	Process concern															x	x				x											x				x		x	x	x			x										x	x	x			
S	Productivity concern																																	x			x								x	x				x			x	x		x	x		
	Flexibility and change		x								x	x				x	x		x	x	x											x		x		x			x				x	x	x	x	x			x	x			x	x	x	
	Job content									x	x				x	x	x			x											x		x	x	x		x	x				x				x		x		x	x		x		x	
	Cycle time	x					x				x	x				x		x	x	x	x											x		x	x			x	x				x			x	x		x		x	x		x			
	Pacing	x					x									x		x													x			x					x			x				x		x				x		x		x	
	Skill levels								x						x					x											x		x	x					x													x					
	Training															x					x											x		x	x	x			x												x		x	x	x	
	Motivation									x	x	x		x	x	x		x	x	x											x		x	x	x			x				x				x					x	x	x	x	x	
	General																																																							
	Employment security	x	x		x																											x	x		x	x		x	x													x	x	x	x	x	
	Overtime policy	x	x	x																													x	x		x													x	x										
	Employee selection			x												x					x											x	x																				x	x	x		
	Recruitment policy																																	x	x																				x	x	x		
	Number of shifts	x	x	x	x		x			x			x				x																		x			x							x	x	x								x		
	Safety								x						x					x											x	x					x															x	x	x		
	Health								x						x					x											x	x																					x	x	x	x	
	Remuneration																																																							
	Payment systems																					x											x	x	x				x	x													x		x			
	Payment structures																					x											x	x					x	x													x	x	x	x	x	
	Pay ranges																																	x	x																					x	x	x	x	x	
	Incentives and rewards																																	x	x				x	x	x													x	x	x	x	x	
	alternatives																																																							
	Implementation												x		x																x		x	x				x	x	x	x	x		x									x	x	x	x		
Q	**Product quality**																																																							
U	Design												x		x		x					x					x					x	x			x			x										x	x	x		x	x	x	x	x
A	Design process												x		x		x					x					x					x	x			x			x										x	x	x		x	x	x	x	x
L	**Process quality**																																																							
I	Capability/inspection								x				x	x	x	x			x	x									x			x			x	x		x	x	x		x						x			x		x	x		
T	Inspection locations								x			x	x	x	x	x		x								x	x	x	x	x					x	x		x	x	x		x						x			x		x	x	x	
Y	inspection frequency														x						x											x	x			x	x		x	x			x										x		x			
	Training												x																					x	x	x		x			x															x	x	x	
	Monitoring								x			x		x	x				x	x											x	x			x			x	x													x		x			
	Total quality																																																							
	Initiatives												x			x																x	x		x	x	x	x		x			x										x		x			
	Documentation														x														x						x	x	x		x														x		x			
	Training												x																					x	x	x	x	x	x	x		x													x		x	x	x	
	Responsiblity																																	x	x		x								x											x	x	x	x	
	Quality levels																																																							
	how select															x																		x			x				x						x			x		x	x		
	levels selected															x																		x		x	x				x						x			x					
	alternatives																																																							

	Supplier relations
P	Inventory
R	Size
O	Spread
D	Balance
	Location
P	Function
	Manuf'g priorities
L	Method
A	Organisational level
N	Centralisation
N	Co-ordination
I	Autonomy
N	Response level
G	Management
	Materials mngmnt
	Customer promises
	Forecasting
	System
	Investment
	Planning
	Time horizon
	Scheduling
	Resource allocation
	Formal paradigms
	Informal
	Centralisation
	Monitoring
	Updating time frame
	Control
	Order release
	Expediting
	Batch sizes
	Prod'n approach
	alternatives
	Product details
N	Scope
E	Focus
W	Range
	Volume
	Introduction
P	Rate
R	Philosophy
O	Product life cycle
D	Computer aids
U	C-A application
C	C-A extent
	Innovation
S	**Lead-times**
	Product design
	Manufacturing

The table below is reproduced from a landscape matrix. Columns are the *Decisions* attributes (grouped: Performance, Structure, State, Management, Functions, Co-ordination); rows are the *Sub-decisions* (grouped: CAPACITY, FACILITIES, PROCESSES, VERTICAL, SUPPLIER, HUMAN RE, QUALITY, PROD PLANNING, NEW PRD, ORGNSAN).

Decisions (column attributes):

Code	Group	Attribute
P	Performance	Selection of criteria
E	Performance	Variables focus
R	Performance	Mgt intg focus
F	Performance	Benchmarking
	Structure	General
O	Structure	Flatness
R	Structure	Formality
G	Structure	Centralisation
A	Structure	Control
N	State	State
I	Management	Openness
S	Management	Product undrstdng
A	Management	Manf undrstdg
T	Management	Systems perspective
I	Management	Culture
O	Functions	Emphasis
N	Functions	Mngmnt superv'n
	Co-ordination	Marketing
	Co-ordination	Engineering
	Co-ordination	Customers

Sub-decisions (row groups and items):

- CAPACITY: Total capacity, Variation Satisfaction, Expansion Methods, Contraction Methods, Timing, Bottlenecks, Demand forecasting
- FACILITIES: Number, Specification, Location, Focus / specialisation, Function Integration, Material Flow, Information flow, Type of Equipment
- PROCESSES: Degree of competitive app, Material handling, Process Organisation, Focus, Man-machine interface
- VERTICAL: Supply chain ownership, Expansion, Contraction, Position in chain, Make vs buy implications
- SUPPLIER: Competitive type, Time span, Sourcing type, Supplier, Products
- HUMAN RE: Cultural properties and cor, Production related, General, Remuneration
- QUALITY: Implementation, Product quality, Process quality, Total quality, Quality levels, Supplier relations
- PROD PLANNING: Inventory, Manufacturing priorities, Management, Forecasting, Planning, Scheduling, Production approach
- NEW PRD: Product details, Introduction, Lead-times
- ORGNSAN: Performance, Structure, State, Management, Functions, Co-ordination

WORKSHEET MSA 4.1.1—Policy/Decision Capture

Project Title:

Person(s) Responsible:

Version: **Date Completed:**

Policy area	Importance	Key decisions and current company practice
Capacity		
Facilities		
Processes and technology		
Vertical integration		
Supplier relations		
Quality systems		
Human resources		
Production planning and control		
New product introduction and scope		
Performance measurement		
Organization		

Task Document MSA 4.2—Current Policy/Practice Analysis

TASK OVERVIEW

DESCRIPTION

Once the responses to MS policy questions are captured, either for individual Product groups or for the entire system, the effects of these on system performance may be assessed. Hence, this task aims to evaluate the company's current practice and identify how this is affecting its competitiveness, and the consistency amongst the captured decisions. This is achieved by assigning an impact value to each of the key decisions previously identified, against each of the competitive criteria. This value indicates both the nature of the impact (either positive or negative, depending on whether the current practice is beneficial or detrimental to the competitive criterion in question) and the degree to which it is affecting performance. Ideally, this assessment should be carried out individually for each Product group. Then, an aggregated assessment can be produced for the MS system as a whole, if necessary. When assessing the current policies, the problem definition table (*Tool/Technique MSA 3.2.1*) provide a means of focusing attention on the more relevant aspects of the strategy policies.

TASK LINKS

INPUT FROM : MSA 4.1

OUTPUT TO : MSA 5.1

POSITION IN MSM FRAMEWORK

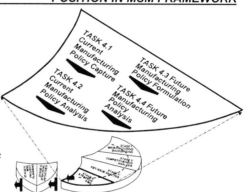

TASK OUTPUTS

Current decision/practice profiles indicating how they are affecting system performance against competitive criteria.

The consistency and coherence of the above also checked.

TASK PROCEDURE

TASK FLOWCHART

	Input	Tool	Output
Step 1	Current company practice, in terms of key decision from previous task.	Wksheet MSA 4.2.1. Tool/Tech. MSA 3.2.1 (Problem definition table).	A collection of completed Wksheet MSA 4.2.1.

For each of the relevant product groups/ system:

* *Assigning an impact value to each of the key decisions previously identified, against each of the competitive criteria.*
* *Record the results in worksheet.*

Use the Problem Definition Table for guidance if required.

All relevant prod. groups/systems considered? — No

Yes

WORKSHEET MSA 4.2.1—Strategic Policy Analysis

Project Title:

Person(s) Responsible:

Version: **Date Completed:**

Type of profile: product-group/system
Product group Name/No (for product group based analysis):

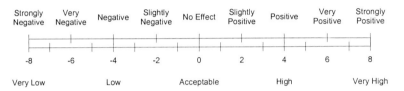

Assessment values for effect of manufacturing policies:

Decision area	Quality	Delivery lead-time	Delivery reliability	Design flexibility	Volume flexibility	Cost
CAPACITY						
Demand Pitch						
Variation satisfaction						
Expansion methods						
Contraction methods						
Timing						
Bottlenecks						
Demand forecasting						
FACILITIES						
Specification						
Location						
Focus						
Function integration						
Flow						
PROCESS						
Type of equipment						
Competitive application						
Material handling						
Process organization						
Focus						
Man-M/c interface						

VERT. INT.								
Supply chain Ownership								
Expansion/ contraction								
Position in chain								
SUPPLIER								
Competitive type								
Time span								
Sourcing								
Supplier qualification								
Partnership								
Make versus buy								
Implications								
HUMAN RES.								
Cultural properties								
Production related								
General								
Remuneration								
Implications								
QUALITY								
Implementation								
Design quality								
Process Quality								
Total quality								
Quality levels								
Implications								
PLANNING								
Supplier & inventory								
Mfg. priority management								
Forecasting								
Planning								
Scheduling								
Control								
Implications								
PRODUCT								
Product details								
Introduction								
Lead-times								
Implications								
PERF. MEAS.								
General								
Implications								
ORGANIZT'N								
Structure								
State								
Management								
Functions								
Coordination								

Task Document MSA 4.3—Future Strategy Formulation

TASK OVERVIEW

DESCRIPTION

In contrast to the previous step, this task actually develops the future strategy with respect to the decisional content of the individual policy areas and records the contents of these decisions in a structured way. The formulation of the future MS strategy follows a similar pattern of questions as used for the capture of the current manufacturing strategic decisions/practice. A number of questions are posed for each of the eleven policy areas to guide the user in formulation and recording particular aspects of the strategy. More importantly, the strategy relationship table, which indicates possible influences and relationships between policy decisions and sub-decisions, can be used to help ensure that a complete strategy is to be formulated and that all the related main issues are addressed.

TASK INSTRUCTION

In relation to the problems identified from the current policy analysis, guidance is provided in the previous stages, such as: the quick-hit table, the associated policy decision relationship table, the problem definition section, the SWOT analysis and the definition of the key issues facing manufacturing. Each of these sections helps the user identify which aspects of the MS function and its strategy need to be addressed. It must be stressed that strategy formulation is a creative process. The decomposition of the eleven policy areas into decisions and sub-decisions is merely presented in order to assist the users to capture and/or develop their own strategies.

An additional aid here is the possible application of the so-called *generic strategies*. While the application of generic strategies on their own has been criticized, they do provide an initial starting point from which to derive a strategic direction and a more detailed specification of the MS strategy. An overview of these are provided as *Tool/Technique MSA 4.3.1*. It is not advisable that they should be applied in their 'pure' format. However, if considered appropriate, they may be tailored and combined with the policy decisions to produce specific strategies. Principally, this involves an assessment of whether the strategy is appropriate for the enterprise, evaluated by assigning it to a phase in the framework and comparing the actual strategy with those considered to be appropriate for the phase in which it resides. In many respects, this stage in the process serves as a means of recording the pattern of actions and decisions that, in practice, are generally conceived over a period of time. The capturing of such policy decisions within the MSA/MSD interface not only allows the pattern to be recorded for particular MSD projects, but also encourages the continual updating of the policy areas. This is necessary due to the dynamic nature of the environment facing all levels of strategy formulation. This implies that formulation and application of effective strategies can only be achieved through repeated iterations, continual review of strategic contents and the inclusion within the process of the evaluation and implementation of *ad hoc* programs and action plans. The MSM structure and, in particular, its MSA/MSD interface presented in this workbook, helps address aspects of these issues by providing a record of the following:
1) The contents of manufacturing strategy at any one time.
2) The state of the company on which the strategy is based, and the resulting action plans and system design tasks.

During the later stages, a means is also provided within the MSM framework to evaluate the effectiveness of the policies both before and after they are actually implemented. As a result, the MSA/MSD interface supports a time-independent process: it allows for reiterations during the strategy capture/formulation process, during the MSD project (when the necessity for a new strategic decision may become apparent), and continuously throughout the MS life cycle.

TASK LINKS

POSITION IN MSM FRAMEWORK

INPUT FROM:	MSA 3.2, 3.3, 3.4
OUTPUT TO:	MSA 4.4

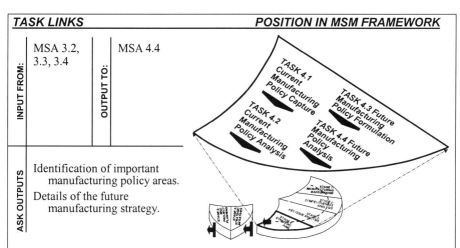

ASK OUTPUTS

Identification of important
   manufacturing policy areas.

Details of the future
   manufacturing strategy.

TASK PROCEDURE

TASK FLOWCHART

	Input	Tool	Output
Step 1	Problem definition results & SWOT analysis.	Worksheet MSA 4.1.1	Identification of key/high-priority policy areas.
Step 2		Tool/Tech. 4.1.1 (Decision list). Tool/Tech. 4.3.1 (Generic strategy profiles).	Initial identification of relevant issues within the key policy areas, and their related future strategic direction.
Step 3	Results from previous step.	Tool/Tech. 4.1.2 (Decision relationship table). Tool/Tech. 4.3.1 (Generic strategy profiles). Worksheet MSA 4.11	Completion of task worksheet: the complete identification of relevant issues, and their related future strategic direction.

TASK FLOWCHART:

Prioritise the eleven policy areas of strategy analysis.

For each of the high priority policy areas:

* *Identify the decisions and sub-decisions that are to be considered.*
* *Formulate future strategic directions for the above.*
* *Record the contents in worksheet.*

For each of (or the key) decisions identified:

* *Cross-check to identify related issues that may also need to be considered.*
* *Formulate future strategic directions for the above.*
* *Record the contents in worksheet.*

Future decisions formulated for all necessary areas? — No

Yes

TOOL/TECHNIQUE 4.3.1—Generic Priority Profiles

Depending on the particular type of MS operation concerned and its current positioning, the ideal strategy is the one that will cause the enterprise to progress towards its goals in a consistent and logical manner. If the state of the enterprise is known, then appropriate strategies may be suggested for consideration. Therefore, the compatibility of MS strategies with respect to the organizational state can be loosely assessed, and alternative types of strategies developed.

One example of a generic strategic priority list classifies manufacturing organizations into distinct types, according to their strategic characteristics, and then prioritizes the competitive criteria shown in the table. Such generic strategies aid in developing a set of generic priority profiles for cross-checking the local requirement profile against general, global expectation.

	Make-for-Stock	**Make-for-Order**
Low Volume	**"Marketeer"** 1. Quality 2. Cost 3. Delivery reliability 4. Delivery lead-time 5. Design flexibility	**"Innovator"** 1. Quality 2. Design flexibility 3. Delivery reliability 4. Delivery lead-time 5. Cost
High Volume	**"Caretaker"** 1. Cost 2. Quality 3. Delivery reliability 4. Delivery lead-time 5. Design flexibility	**"Reorganizer"** 1. Delivery reliability 2. Delivery lead-time 3. Quality 4. Cost 5. Design flexibility

These profiles are not provided to the companies in a prescriptive manner, but only as suggestions for exploring their own strategic approach. Hence, by considering the corporate and business strategies, the competitive criteria analysis, key issues, SWOT analysis results and problem definitions, a generic approach can be customized and then used to specify future MS policy decisions.

By comparing the generic strategy profiles with those based on the results of a firm's own analysis, the strategy formulation also takes into account the development of competitive criteria, capabilities and competencies. It also considers the global expectation, to certain extent. If the user consider this a useful guide, the generic priority profiles shown may be used as a reference during this stage of the analysis.

Task Document MSA 4.4—Future Strategy Analysis

TASK OVERVIEW

DESCRIPTION

The future MS strategy analysis serves to consolidate the strategy formulation process. Since the many decisions within the policy areas are interdependent, a degree of trade-off is likely to be required, both at the level of the competitive criteria and at the level of individual decisions. Having to make such compromises makes it necessary to assess the overall effects. This ensure that the strategy is still aimed in the same direction, and that the policies formulated are consistent and coherent. Consequently, this task aims to provide an initial indication of how the manufacturing strategy, as defined within the individual policy areas, may affect the future competitiveness of the MS function. It employs the same worksheet used for current policy analysis to assess the principal policies with respect to their effect on the competitive criteria. Since the analysis can only be subjective at this stage, discussion of the validity of the assigned values and of the results of the analysis is encouraged, and iterations between this task and the previous tasks are expected.

TASK LINKS   POSITION IN MSM FRAMEWORK

INPUT FROM : MSA 3.4, 4.1, 4.2, 4.3.

OUTPUT TO : MSA 5.1

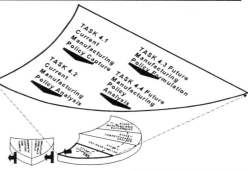

OUTPUTS

Future decisions, consistency checked to ensure they will affect system performance in the desired direction.

TASK PROCEDURE   TASK FLOWCHART

	Input	Tool	Output
Step 1	Initial future policies, in terms of key decisions from the strategy formulation task.	Wksheet MSA 4.2.1.	
Step 2	Results from SWOT analysis. Key issue statements	Tool/Tech MSA 3.2.1 (Problem definition table).	Several completed Wksheet MSA 4.2.1

For each of the relevant product groups/system:
* *Assigning an impact value to each of the key decisions previously identified, against each of the competitive criteria.*
* *If required, calculate utility value to give indication about the decisions' overall effects on criteria.*

Effects of decisions consistent with requirements? —No→ Readjust policy contents to achieve consistency

All relevant prod. groups/systems considered? —No

Yes

CHAPTER THREE

MS Strategy and System Design Interfacing

3.1 INTRODUCTION

This is the point at which, through a number of iterations, the process seeks to find the appropriate action plans on the MS functions. The end result is to support and provide a competitive advantage. As shown in Figure 3.1, it consists of two stages:

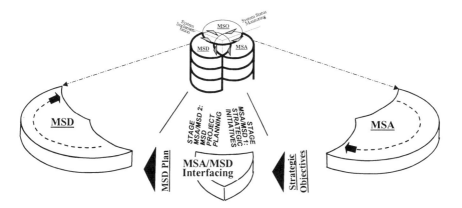

Figure 3.1 Interfacing between MSA and MSD

- **Stage MSA/MSD 1—Strategic Initiatives**. This stage defines how the strategic aims and MS policies specified in the previous stage will be achieved. It therefore represents the first steps of the MSA to MSD linking process. The key element of this stage is the development of action plans through which the company can attempt to implement the required strategies and policies.
- **Stage MSA/MSD 2—MSD Project Plans**. This is the second stage of the MSA/MSD linking process. It involves the refinement of the action plans to specify particular MSD project(s). The project terms-of-reference are defined before the project itself is specified in terms of its constituent MSD tasks, together with their aims, targets and constraints.

3.2 STAGE MSA/MSD 1—STRATEGY INITIATIVES

As shown in Figure 3.2, this stage consists of two tasks. The first task involves the identification of the main changes in MS strategy policies and decision areas. The second task specifies action plans that implement the required changes. Again, the tasks involve the completion of a series of tables and the selection and development of appropriate action plans based on the results previously generated through the MSA/MSD process. When applied to the example company, this stage produces the following results:

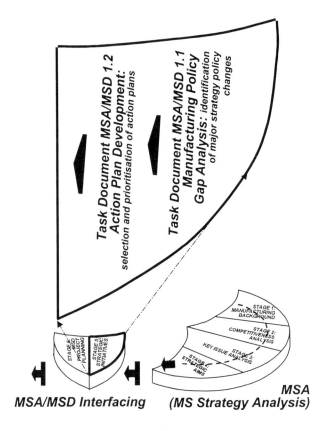

Figure 3.2 Stage MSA/MSD 1—strategy initiatives

MSA/MSD 1.1—Policy Gap Analysis (Worksheet MSA/MSD 1.1.2)
Within the example, no major changes can be immediately identified. However, technology adoption is to be more restrained and focused, and a capacity increase is to be achieved through the development of a new site and used to provide a focus for the site, as shown in Table 3.1.

Table 3.1 Example—statement of strategic initiatives

Key Decision Areas		Description of Strategic Aims	Effect on Competitive Criteria					
			Quality	Lead-time	Delivery reliability	Design flexibility	Volume flexibility	Cost
1	*Capacity*	Increase capacity through new equipment and facility.		+	+		+	-
2	*Facilities*	New site development, adopt cellular manufacture where beneficial, try to simplify material flow, split site between core businesses.		+	+		+	
3	*Processes & Technology*	Apply technology only for the benefits, adopt standard modular machine tools rather than expensive flexible machine tools.				-	-	+
4	*Supplier Development*	Change policy to farm out volume bits to subcontractors.				-	-	+
5	*Human Resources*	Develop job skills, increase quality concern.	+	+	+		+	+
6	*Quality Systems*	Continuous improvement of quality program, SPC, quality circles, in-process inspection, ISO 9000.	+	+	+			+
7	*Planning and Control*	Reduce inventory, improve control, simplify material flow, improve capacity planning required.		+	+		+	
8	*Scope & New Products*	QFD, concurrent engineering.	+			+	+	
		Overall Effects	+3	+5	+5	-1	+3	+3

MSA/MSD 5.2—Action Plan Development

Based on the direction of the future strategy and the competitive requirements, the following action plans are selected and recorded in *Worksheet MSA/MSD 1.2.1*:

- capacity expansion,
- relocation and focusing of facilities,
- equipment improvement,
- workforce development,
- order-to-Delivery lead-time reduction,
- setup time reduction.

Task MSA/MSD 1.1—Policy Gap Analysis and Initiatives

TASK OVERVIEW

DESCRIPTION

A gap analysis is used to highlight specific directions for change in order to assist in the development of MSD action plans. Hence, it is the differences between the current and future policies that need to be identified. There are two stages to the gap analysis. The first stage simply identifies those questions that have been answered within the policy area questionnaire, and those questions that have different responses. The second stage carries out a gap analysis of the strategy assessments to ascertain the extent of the effects on the competitive criteria of the new strategy in comparison to the old one. Gap analysis is applied separately to each Product group, or to the system, as a comparison of the old and new, producing a series of policy-criteria gap matrices as an output. Additionally, where the responses are suitably scaled for a radar representation, the policy analysis can be directly compared to the market requirements profiles of Stage MSA 2. Through such comparison, current and future policies may both be seen to contribute towards the attainment of market requirements.

TASK LINKS   POSITION IN MSM FRAMEWORK

INPUT FROM:	MSA 4.1, 4.2, 4.3, 4.4	OUTPUT TO:	MSA/MSD 1.2

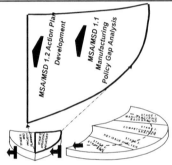

OUTPUTS	Identification of major policy changes, possible effects of which are checked against market requirements.

TASK PROCEDURE   TASK FLOWCHART

	Input	Tool	Output
Step 1	Current and future policies, in terms of key decisions from the strategy policy questionnaire	Wksheet MSA /MSD 1.1.1	
Step 1	Market requirements Results from SWOT analysis Key issue statements	Tool/Tech. MSA 3.2.1 (Problem definition table)	Several completed Wksheet MSA/MSD 1.1.1

For each of the relevant product groups/system:
* *Identify the key questions that have been answered within the questionnaire.*
* *Identify the questions that have generated different responses from the current practice and the future strategy.*
* *If required, calculate utility value to give indication about the effects of these differences on competitive criteria.*

Effects of differences consistent with market requirements? —No→ Readjust policy contents to satisfy market requirements.

All relevant prod. groups/systems considered? —No→

Yes

WORKSHEET MSA/MSD 1.1.1—Policy Gap Analysis & Initiatives

Project Title:

Person(s) Responsible:

Version: **Date Completed:**

Type of profile: product-group/system
Product group Name/No (for product group based analysis):

Decision Area	Quality		Delivery lead-time		Delivery reliability		Design flexibility		Volume flexibility		Cost	
CAPACITY												
Demand Pitch												
Variant satisfaction												
Expansion methods												
Contraction methods												
Timing												
Bottlenecks												
Demand forecasting												
FACILITIES												
Specification												
Location												
Focus												
Function integration												
Flow												
PROCESS												
Type of equipment												
Competitive application												
Material handling												
Process organization												
Focus												
Man-M/C interface												
VERT. INT.												
Supply chain ownership												
Expans'n/cntrac'n												
Position in Chain												

SUPPLIER								
Competitive type								
Time span								
Sourcing								
Supplier qualification								
Partnership								
Make versus buy								
Implications								
HUMAN RES.								
Cultural properties								
Production related								
General								
Remuneration								
Implications								
QUALITY								
Implementation								
Design quality								
Process quality								
Total quality								
Quality levels								
Implications								
PLANNING								
Supplier & inventory								
Manufacturing priority								
Forecasting								
Planning								
Scheduling								
Control								
PRODUCT								
Product details								
Introduction								
Lead-times								
PERF. MEAS.								
General								
Implications								
ORGANIZATN								
Structure								
State								
Management								
Functions								
Coordination								

WORKSHEET MSA/MSD 1.1.2—Statement of Strategic Initiatives

Project Title:

Person(s) Responsible:

Version: **Date Completed:**

Type of profile: product-group/system
Product group Name/No (for product group based analysis):

Key Decision Areas	Description of Strategic Aims	Quality	Lead-time	Delivery reliability	Design flexibility	Volume flexibility	Cost
		\multicolumn{6}{}{**Effect on Criteria** ("-": negative/"+": positive)}					
1							
2							
3							
4							
5							
6							
7							
8							
	Overall Effects						

Task MSA/MSD 1.2—Identification of Action Plans

TASK OVERVIEW

DESCRIPTION

This task aims to identify action programs to assist the implementation of the formulated manufacturing strategies and policies. The key inputs for the selection process are the main changes required between the current and future manufacturing strategies. These provide a list of strategic initiatives, because they are essentially a statement about how the future strategic aims are to be achieved. Hence they form the basis for a number of MS system objectives and action plans, either in a particular area, or across a number of departments.

In order to assist in the specification of operating plans, a table of generic action plans is provided. These plans represent an aggregation of those identified in the literature and those observed in industrial practice from case studies. They provide a broad cross-section of the types of MSD actions likely to be required, and range from complete MSD projects to continuous improvement programs. In the table, they have been grouped approximately according to their fit within the manufacturing policy areas and MSD task frames. When applied in combination with the strategic initiatives from the previous steps, an indication of the prospective operating plans can be produced, forming the foundation for the specification of MSD project terms-of-reference in the next stage of the MSA/MSD interfacing process.

TASK LINKS   POSITION IN MSM FRAMEWORK

INPUT FROM : MSA/MSD 1.1

OUTPUT TO : MSA/MSD 2.1

OUTPUTS Identification of relevant action plans.

TASK PROCEDURE   TASK FLOWCHART

Step 1

Input	Tool	Output
Key issues Major policy changes	Wksheet MSA /MSD 1.2.1	Several completed Wksheet 1.2.1

For each of the relevant product groups/ system:

* For each of the major policy changes needed, identify the most relevant action plans from the generic list.
* Assign a value (1-10) to indicate its priority.
* Record the results in worksheet.

All relevant prod. groups/system considered? — No

Yes

WORKSHEET MSA/MSD 1.2.1—Action Plan Identification

Project Title:

Person(s) Responsible:

Version: **Date Completed:**

Type of profile: product-group/system
Product group Name/No (for product group based analysis):

Action Plans	Priority	Action Plans	Priority
Strategy		**Planning and Control**	
Link to business strategy		Production/inventory control systems	
Define manufacturing strategy		Production/inventory control systems training	
Activity-based costing		Just in time manufacture	
Capacity and Facilities		Supplier lead-time reduction	
Increase capacity		Reduce provisioning time	
Lead-time reduction		**Quality Systems**	
Reduce setup times		Establish total quality control program	
Focus factories		Zero defects	
Manufacturing reorganization		Statistical process control	
Group technology		Quality function deployment	
Improve existing systems		Statistical quality control	
Recondition existing plants		Quality circles	
Relocate plant		Improve suppliers quality	
Close plant		Preventative maintenance	
Processes and Technology		Improved maintenance	
New process, old product		**Vertical Integration**	
New process, new product		Optimize "make versus buy" mix	
Improve equipment and process technology		Improve distribution	
Improve energy/utilities efficiency		**Human Resources**	
Reduce materials losses		Direct personnel training	
Improve equipment utilization		Supervisory training	
Increase operations standardization		Manufacturing management education	
Manufacturing mechanization		Reduce lost work time	
Introduce FMS		New wage system	
Introduce robots		Direct labor motivation	
Introduce material handling		Apply rewards and penalties	
Introduce CAM		Productivity bargaining	
Introduce CAD		Employee productivity gains-sharing	
Increase technical autonomy		Redesign jobs	
Automate operations		Specialize jobs	
Product Scope and New Products		Broad scope of work	
Narrow product lines/standardization		Involve workers in planning	
Reduce number of variants		Broad planning responsibility	
Redesign of products		Ergonomics	
Value analysis/product design		Worker safety	
Design for manufacture		Reduce number of employees	
Develop product workshops		New skills hiring	
Product introduction ability improvement		Develop workforce with multiple/flexible skills	
Information Systems		Improve work methods and procedures	
Manufacturing information systems		Implement group work	
Integrated manufacturing information systems		Inter-functional work teams	
Inter-functional information systems		**Organization**	
Integrated information systems		Change management/management relations	
Office automation		Encourage employee involvement	
Decentralize decision-making authority		Improve departmental performance	
Improve information handling		Change organizational design/focus	
Improve communications		Improve integration among departments/functions	
Building			
Work environment improvement			
External environment improvement			

3.3 STAGE MSA/MSD 2—MSD PROJECT PLANNING

This stage completes the process of MSA/MSD interfacing. Associating future strategic requirements with MSD tasks helps the analyst identify the relevant MSD tasks and layout the project execution and system implementation plans.

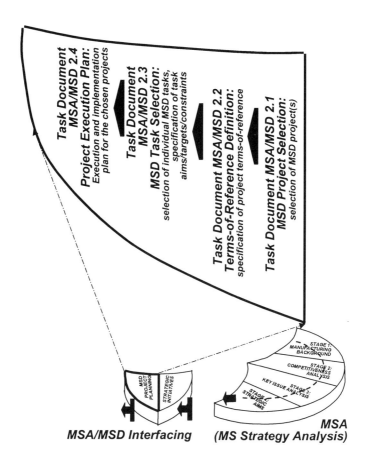

Figure 3.3 Stage MSA/MSD 2—MSD Project Planning

As shown in Figure 3.3, it consists of four task documents. The first involves ranking and weighting the previously selected action plans. This is achieved by responding to a number of questions based on the results previously generated, and on the intended MSD project to be undertaken. The second task helps to specify detailed terms-of-reference for the planned MSD project(s) in terms of objectives, targets and constraints. The third task involves the use of a number of linking tables to assist the selection of the appropriate MSD tasks. This particular task is a slightly more complex process than most. However, detailed instructions and other aids provided in the task document should make its completion a structured and straightforward process.

Finally, plans are produced for the execution of the design projects and the implementation of the new system. When applied to the example company, this analysis produces the following results:

- *Action Plan Selection*—From the initial set of action plans chosen to implement the new manufacturing strategy, a selection of action plans are grouped using *Worksheet MSA/MSD 1.2.1* to form the basis for the MSD project. For the purpose of illustration, only one overall project is assumed here. This project includes: (1) capacity expansion, (2) relocation, (3) reduce order to Delivery lead-time and (4) reduce setup times.
- *Terms of Reference Definition*—Once the MSD project is defined with respect to the action plans it is aiming to implement, the next stage is to define the project's terms of reference, particularly the project scope and objectives, using *Worksheet MSA/MSD 2.2.1*.
- *MSD Project Scope*—Existing product, existing system, redesign, physical system, factory to workstation levels (though predominantly product unit to workstation levels), initiated by business requirements, solutions driven by cells and JIT philosophies.
- *MSD Project Objectives*—Reduce production costs, reduce lead-times, increase throughput, increase Volume flexibility, increase production volume, reduce non-value-adding activities, and simplify material flow.
- *Task Selection*—Following *Task MSA/MSD 2.3*, several sets of task selections can be generated for the example company. The utility values and/or the subsequent percentage values for each set of relationship tables indicate, in an approximate fashion, the degree of relevance of each task to the rationale behind their selection. The first-pass MSD tasks thus suggested are as summarized in Table 3.2.

Table 3.2 Example—initial MSD project plan and detailed MSD tasks

Task Frame Order	Major Tasks	Secondary Tasks	Additional Tasks
System Function	Process, Analysis Make versus Buy (1)	Product Analysis Part Analysis	-
System Structure	Capacity Demand	Functional Grouping Structural Layout Integration-Modularization	-
System Decisions	-	-	Information Functions Decision Variables
Physical System	Make versus Buy (2) Conceptual Capacity	Process Planning Part Grouping Cell Formation Conceptual Layout Material Handling Factory storage Support Facilities	Space Determination
Organizational System	-	Organization Structure Labor Policy Quality Policy	Organization Culture Organization State
Information System	-	Planning and Control	Integration System Architecture Data Flows
Manufacturing	Equipment Selection	Domain Location/Layout Detailed Cell Layout Workstation Layout	-
Logistics	-	Storage Location Storage System	Buffer Sizes Handling Path Handling Unit
Support	-	Maintenance Tooling Supplies Setup Management Process Inspection	Administration
Building and Facilities	-	Machine Services	Human Services Material Services Building
Planning	-	Production Planning Scheduling Batch Sizes Volume Mix	Shift Patterns
Control	-	Control Systems Materials Management	Data Collection
Human	-	Job Requirements Job Design	Training Quality

Task MSA/MSD 2.1—MSD Project Selection

TASK OVERVIEW

DESCRIPTION

Based on the strategic requirements and related action plans previously identified, the aim of this task is to identify clearly the range of MSD projects required. Depending on the number of action plans derived and the opinions of management and the design team, the MSD process will either take the form of a large all-encompassing project or a series of individual projects. In the latter case, a means is required to guide the selection and prioritization of the appropriate action plans. Even in the former case, a ranking and weighting exercise should be carried out to prioritize the action plans within a large MSD project.

Factors to be considered include: objectives, sites to be affected, time span, availability of skills and resources, disruption caused, financial implications, cost estimates, group consensus, the logical sequencing and activity dependencies of action plans.

TASK LINKS

INPUT FROM : MSA/MSD 1.4

OUTPUT TO : MSA/MSD 2.2

OUTPUTS MSD project(s) consisting of previously identified action plans

POSITION IN MSM FRAMEWORK

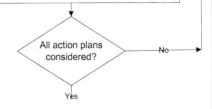

Task Document MSA/MSD 6.4
Project Execution Plan:
Execution and implementation plan for the chosen projects

Task Document MSA/MSD 6.3
MSD Task Selection:
selection of individual MSD tasks, specification of task aims/targets/constraints

Task Document MSA/MSD 6.2
Terms-of-Reference Definition:
specification of project terms-of-reference

Task Document MSA/MSD 6.1
MSD Project Section:
selection of MSD project(s)

M S A

TASK PROCEDURE

TASK FLOWCHART

	Input	Tool	Output
Step 1	List of chosen action plans	Tool/Tech. MSA/MSD 2.1.1 (check list)	
Step 2		Tool/Tech. MSA/MSD 2.1.1 Wksheet MSA /MSD 2.1.1	Completed Wksheet 2.1.1

Decide number of MSD projects needed, considering factors such as objectives, time span, availability of skills and resources, disruption caused, financial implications, etc.

For each of the action plans listed before:

* *Assign to appropriate MSD project*
* *Record its priority value*

All action plans considered? — No

Yes

TOOL/TECHNIQUE MSA/MSD 2.1.1—Project Check-list

The following checklists assist in project selection and definition of project term-of-reference.

How much of the organization will be affected by the project? This determines the scope of the project in terms of employees who will be directly involved in the changes, and identifies how much training and resources will be required:

- totally within the function,
- between two or more functions,
- within a single process,
- totally within the organization,
- between two or more organizations.

How many locations will be affected by the project? This identifies the issues relevant to the MS units of a distributed MS system located in different sites/regions/countries:

- only one site,
- between two or more sites,
- within one region,
- between two or more regions,
- within a single country,
- between two or more countries,
- every system location.

What is the degree of change for the business as a result of the project? This identifies how much change the company will need to go through for the next year and how many resources will be required:

- totally within the function,
- no change,
- minor change,
- significant change,
- radical.

Assessment of the project's role concerning functional factor. This identifies what functional factors play the most important role for the project and which factors to focus on:

- dependence of the business functions or processes on the system,
- effectiveness in supporting the business functions/processes,
- performance under service-level agreements,
- costs of operation and maintenance,
- backlog of changes request,
- integration of data with other systems.

WORKSHEET MSA/MSD 2.1.1—MSD Project Formulation

Project Title:

Person(s) Responsible:

Version: **Date Completed:**

The chosen action plans can be grouped into short, medium and long-term projects with reference to strategic requirements, and/or according to functional or geographical areas.

MSD PROJECT	ACTION PLAN	PRIORITY
Site: 1		
Site: 2		
Site: 3		
Site: 4		

Task MSA/MSD 2.2—Definition of Terms-of-Reference

TASK OVERVIEW

DESCRIPTION	Based on the strategy contents previously captured, this task aims to refine the future business and MS goals. In addition to setting clearly defined objectives and constraints for the MSD projects, it contributes towards the next stage of the MSA/MSD linking process—that of the identification of the necessary MSD tasks, themselves.
TASK INSTRUCTIONS	In order to derive a set of core project objectives, the strategic objectives identified through Stages MSA 3, 4 and MSA/MSD 1, are to be reassessed and prioritized in consideration of the action plans to be executed. Generally speaking, these should be based on the following strategic drivers for change: new product introduction, capacity adjustment, cost improvements, lead-time reduction, quality improvement, delivery performance and increased flexibility of the MS system. As a result, the following should be defined: *Project Scope*: the product-system relationship, the type of project, the focus of the project, the corporate level of abstraction of the project, the initiators of the project and possible solutions. *Project Objectives*: cost related—reduce production cost (direct/indirect labor, overheads); operations related—reduce lead-time, improve schedule adherence, increase throughput, increase productivity, reduce inventory level, reduce work in progress, reduce non-value-adding activities, increase level of automation, increase process flexibility, increase Volume flexibility, increase production volume, increase integration; organizational restructuring related—simplify material flow, rationalize product range, standardization, increase decentralization, simplify control system, improve ownership; quality related—improve quality, reduce waste, achieve zero defects. *Project and System Constraints*: time constraints, resource constraints, human resource constraints and financial constraints. The list of performance measures identified in the MS performance measure (MPM) stage will provide the overall guidance for this task. In addition, to assist the objective definition, an objectives catalog/objectives tree is provided in *Tool/Technique MSA/MSD 2.2.1*. The selected objectives can then be assessed according to their degree of desirability, feasibility or probable rate of successful achievement, as well as expected resources required. It should also be noted that, in a domain as complex as an MS system, there are likely to be conflicts. Any conflicting terms of reference, particularly with respect to objectives, need to be resolved through three approaches: (1) the short term approach of trading-off objectives, (2) improving the performance of one objective at the expense of another, (3) the optimal solution of moving the 'pivot' of the trade-off, such that long-term improvement of all aspects of performance is possible, as shown in the figure below.

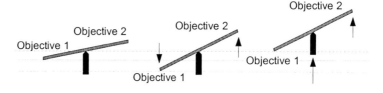

a) Trade-off between (1) and (2).   **b)** Moving the pivot.

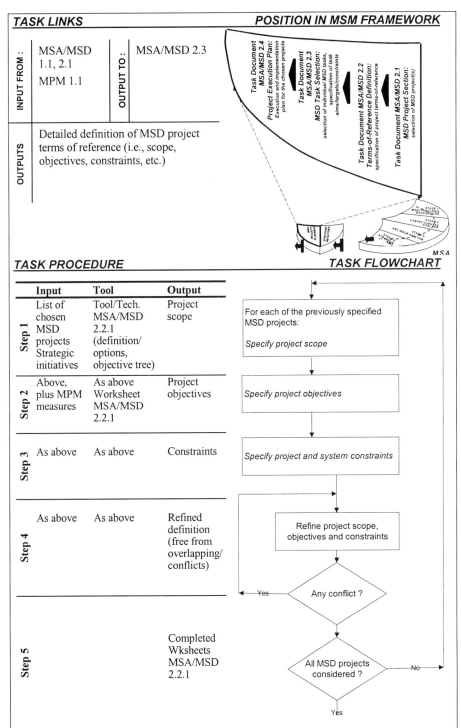

TASK LINKS

INPUT FROM :	MSA/MSD 1.1, 2.1 MPM 1.1	OUTPUT TO ..	MSA/MSD 2.3

OUTPUTS	Detailed definition of MSD project terms of reference (i.e., scope, objectives, constraints, etc.)

POSITION IN MSM FRAMEWORK

Task Document MSA/MSD 2.4
Project Execution Plan:
Execution and implementation plan for the chosen projects

Task Document MSA/MSD 2.3
MSD Task Selection:
selection of individual MSD tasks, aims/targets/constraints

Task Document MSA/MSD 2.2
Terms-of-Reference Definition:
specification of project terms-of-reference

Task Document MSA/MSD 2.1
MSD Project Section:
selection of MSD project(s)

M S A

TASK PROCEDURE

	Input	Tool	Output
Step 1	List of chosen MSD projects Strategic initiatives	Tool/Tech. MSA/MSD 2.2.1 (definition/ options, objective tree)	Project scope
Step 2	Above, plus MPM measures	As above Worksheet MSA/MSD 2.2.1	Project objectives
Step 3	As above	As above	Constraints
Step 4	As above	As above	Refined definition (free from overlapping/ conflicts)
Step 5			Completed Wksheets MSA/MSD 2.2.1

TASK FLOWCHART

For each of the previously specified MSD projects:

Specify project scope

Specify project objectives

Specify project and system constraints

Refine project scope, objectives and constraints

Any conflict ? —Yes→

All MSD projects considered ? —No→

Yes

TOOL/TECHNIQUE MSA/MSD 2.2.1—Objective Trees

Category	Definition (Project Scope)	Option
Product System Relationship	The relationship between the intended products and the intended manufacturing system.	Existing products—existing system, New & existing products—existing system, New products—existing system, Existing products—new system, New & existing products—new system, New products—new system.
Type of Project	The type of project and where it appears in the system life cycle.	System redesign, step improvement, system optimization, new system design, continuous improvement, new manufacturing philosophy.
Focus of Project	The manufacturing sub-systems being addressed by the project, i.e., defining the boundaries of the project on a technical level.	Physical sub-system, control sub-system, decisional sub-system, support sub-system, organizational sub-system, system integration.
Corporate Level	The level that the MSD project is being targeted in terms of systems restructuring and development.	Corporate level, factory level, product unit/module level, cell/workstation level.
Initiators of Project	The drivers of the MSD project with respect to the reasons why the project is to be started.	Adoption of a manufacturing philosophy (typically cells, JIT, TQM, Distributed Manufacturing Systems, etc.), adoption of a business philosophy (BPR, Virtual Enterprises, etc.), new products requiring manufacturing facilities (design and market driven), new strategic business requirements, problems with the existing system.
Possible Solutions	Whether the project involves a possible "solution" which has already been decided by higher management, consultants, etc.	Introduction of cellular manufacturing, introduction of JIT, introduction of CIM, introduction of group technology, introduction of TQM.
Category	**Definition (project constraints)**	
Time Constraints	Project time constraints which are be considered within process planning are basically those employed as project management issues, typically the time constraint on the overall project and the dates of the major review periods, project milestones and gateways.	
Resource Constraints	The resource constraints are concerned with the availability of suitable tools, techniques and facilities, including work space, computers, design equipment, etc.	
Human Resource Constraints	The human resource constraints cover the availability of suitable Task Force Team members for particular design tasks. It also includes other company employees and members of staff and MSD consultants. In particular the constraints involve the experience, skills and knowledge of the members to be assigned to design tasks, and whether these are to be on a full-time or part-time basis.	
Financial Constraints	The project financial constraints involve the running costs of the project with respect to salaries and resource costs. It is also linked to the manufacturing system financial constraints via the capital spending budget for the project.	
Category	**Definition (System Constraints)**	
Time Constraints	The system time constraints cover the operational constraints on the manufacturing system, for example, the duration of the working day, working week and annual holidays. Other constraints, such as the desired throughput time for particular products, can be considered as fixed project objectives, and hence will be assessed under the project objectives section of the terms of reference.	
Resource Constraints	The system resource constraints consider the constraining factors of the manufacturing system with respect to working area and facilities and the capacity and availability of machines and equipment.	
Human Resource	Human resource constraints involve the number, availability, experience, skills and knowledge of the work force.	
Financial Constraints	The financial constraints cover the capital cost of new system equipment and the various specific operational costs of the manufacturing system. In addition, there may be particular constraints on the costs of producing specific products and parts or using specific processes.	

WORKSHEET MSA/MSD 2.2.1—Terms of Reference

Project Title:

Person(s) Responsible:

Version: **Date Completed:**

MSD project title:

Project Scope
Product/System Relationship

	Existing system	New system
Existing products		
New and existing products		
New products		

Type of Project

System redesign		Step improvement	
System optimization		Continuous improvement	
New system design		New manufacturing philosophy	

Project Focus

Physical systems		Support systems	
Control systems		Organizational systems	
Decisional systems		Systems integration	

Corporate Level

Corporate		Product unit/module	
Factory		Cell/workstation	

Project initiators

Adopt manufacturing philosophy	e.g. cells, JIT, TQM, etc.
Adopt business philosophy	e.g. following BPR techniques
New products requiring new facilities	design and market driven
New strategic business requirements	strategy driven
Existing manufacturing system problems	problem driven

Possible Solutions to Adopt

Introduce group technology		Introduce TQM procedures	
Introduce cellular manufacturing		Introduce JIT manufacturing	
Introduce CIM		Other.........	

Project Objectives

Cost Related

Reduce production costs		Reduce materials costs	
Reduce direct/ indirect costs		Other.........	
Reduce overheads			

Operations Related

Reduce lead-times		Increase production volume	
Increase productivity		Increase throughput	
Reduce NVA activity		Reduce work in progress	
Increase Volume flexibility		Increase process flexibility	
Improve schedule adherence		Increase integration	
Reduce inventory		Other.........	
Increase automation			

Organization Related

Simplify flow		Simplify control system	
Rationalize product range		Improve ownership	
Standardize/ commonality		Increase decentralization	
Other			

Quality Related

Improve quality	
Reduce waste	
Achieve zero defects	

Project
Constraint

Time Constraints

Resource Constraints

Human Resource Constraints

Financial Constraints

System
Constraints

Time Constraints

Resource Constraints

Human Resource Constraints

Financial Constraints

Objectives Refinement

Objectives	Desirability	Feasibility	Availability	Priority	Targets

Task MSA/MSD 2.3—Selection of MSD Tasks

TASK OVERVIEW

Once a relevant project is identified, it is then necessary to specify the individual tasks that need to be carried out within it. For this purpose, the action plans previously listed can be used to highlight relevant tasks. Following this, both the MS strategy and its MSD terms-of-reference can provide a further indication of the emphasis of the particular MSD project(s) under consideration, and hence, a means of refining task selection. The refinement can be in terms of task choice and task definition, such as individual task aims, scope, target and constraints. Tables are provided to assist in the selection process:

Generic action plans—MSD tasks linking (Tool/Technique MSA/MSD 2.3.1). This table provides an indication of the possible relationships between each of the generic action plans and the MSD tasks (relationships between the 75 MSD tasks and the 88 generic action plans).

Strategic decisions—MSD task linking (Tool/Technique MSA/MSD 2.3.2). This provides an indication of the possible relationships between each of the sub-decisions, categorized under the strategic decisions of each of the manufacturing policy areas, and the generic MSD tasks (currently over 200 separate sub-decisions grouped under 55 decisions within the 11 policy areas). When this table is analyzed it can be seen that the mapping of policy areas to sub-systems, as outlined in Chapter 1, though simplistic in nature, corresponds sufficiently with the more detailed level of abstraction.

Project terms-of-reference—MSD tasks frames (Tool/Technique MSA/MSD 2.3.3). This provides an indication of the general relationships between each of the project terms of reference with respect to the project scope, objectives, and the MSD tasks.

By utilizing these tables, the top-down approach to linking MS strategy to the MSD process can be followed through three stages:

Task identification. As the preliminary step, this produces a list of candidate tasks.

Task verification. The preliminary choice of MSD tasks needs to be reduced and/or refined by referring to the original strategic requirements. This helps prioritize the task choices and assists in making the final choice.

Task definition. The overall project criteria and terms-of-reference need to be disaggregated to set aims, targets and constraints for chosen individual tasks, hence making sure the individual task objectives are consistent on a global level.

TASK LINKS   *POSITION IN MSM FRAMEWORK*

INPUT FROM :		OUTPUT TO :
MSA/MSD 2.1 (project list)		MSD task modules as identified.
MSA/MSD 1.1 (strategy initiatives)		Core MSM functional area.
MSA/MSD 2.2 (term-of-ref.)		MSM status monitoring area.
MSA/MSD 3.4 (key issues)		

OUTPUTS

List of MSD tasks for each necessary project, covering requirement specification, conceptual and detailed design stages. Detailed objectives/aims specified for individual tasks.

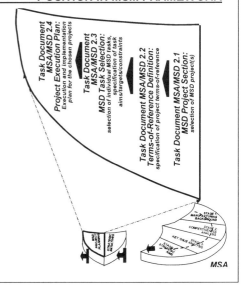

TASK PROCEDURE

TASK FLOWCHART

	Input	Tool	Output
Step 1	MSD project list and associated action plans (Wksheet MSA/MSD 2.1.1).	Action Plan/MSD Task linking table (Tool/Tech. MSA/MSD 2.3.1).	Preliminary list of MSD tasks.
Step 2	List of preliminary MSD tasks. Strategic initiatives (Worksheet MSA/MSD 1 1.1).	Decisions/MSD Task linking table (Tool/Tech. MSA/MSD 2.3.2).	Final choice of MSD tasks for each project.
Step 3	List of final MSD tasks. Project term-of-reference (Wksheet MSA/MSD 2.2.1). Strategic Key Issues (Wksheet 1.1.1).	Term-of-Ref/MSD Task linking table (Tool/Tech. MSA/MSD 2.3.3).	Task level objectives, targets and constraints.

For each of the intended MSD projects in Worksheet MSA/MSD 2.1.1:

TASK IDENTIFICATION FROM ACTION PLANS
For each of the action plans listed for the project:
* Locate the particular action plan in the Action Plan/MSD Task Linking Table (Tool/Tech. MSA/MSD 2.3.1).
* Cross check to identify the MSD tasks related to this action plan.
* From this list, make a preliminary choice of tasks for this particular project.
* Record all the identified MSD tasks in Worksheet MSA/MSD 2.3.1, according to their position within the MSD taskframe.

TASK VERIFICATION FROM STRATEGY INITIATIVES
For each of the preliminary MSD tasks:
* Locate the particular task in the Decision/MSD Task Linking Table (Tool/Tech MSA/MSD 2.3.2).
* Cross check to identify the strategic/decision issues related to this task.
* Check the relevance of above against key strategic initiatives as identified in Wkshet MSA/MSD 1.1.1.
* Eliminate irrelevant or less significant tasks from the list. Add new ones if necessary.

All tasks verified? — No

TASK SPECIFICATION
For each of the verified MSD tasks:
* Locate the particular task in the Term of Reference/Task Linking Table (Tool/Tech MSA/MSD 2.3.3).
* Cross check to identify the term-of-reference items related to this task.
* Refer to the overall project requirements as listed in the term-of-reference (Wkshet MSA/MSD 2.2.1), and specify detailed objectives, aims, targets, and constrains for this particular MSD task. Refer to the key issues (Worksheet MSA 3.4.1) and the key strategic initiatives (Wkshet MSA/MSD 1.1.1) if necessary.
* Record results in Worksheet MSA/MSD 2.3.2.

All projects considered? — No

TOOL/TECHNIQUE MSA/MSD 2.3—MSD Task Verification

So far as the MSA/MSD linking process is concerned, the following should be pointed out:
(1) The contents of the linking tables may be edited by the users to present their specific
knowledge, reflect their past experience and match their particular strategic requirements;
(2) The final part of planning the design process, the design task refinement stage, is in
effect the first stage of the design task execution. It involves the configuring of the design
task with respect to its specific objectives, activities to be carried out, and tools/techniques
to be applied.

Since the tasks suggested by the current MSM framework act only as guidelines, the
user may accept the suggested design tasks, edit the selection or pick a totally different set of
tasks. Because every MSD case is different, and because the choice of the next task to
complete is often dependent on the outcome and data obtained for the current tasks, it is
expected that users often return to the task selection stage to change their original choice. In
other cases, it may be necessary to return to the task selection step during the execution of a
task, due to the absence of a critical piece of information that can only be derived from a
task which has not been selected or completed. Again, users have the option of returning to
the task selection stage if they so wish. In both cases, however, it is essential that the
reasons, assumptions and rationale behind the decision be recorded.

Due to the manner in which they are constructed, the table entries are essentially
linear in nature. Since the MS strategy process may produce multiple action programs and
MSD tasks, there is a need to take into account this multi-faceted aspect. The use of multiple
criteria, through ranking and utility values, may provide a suitable approach:

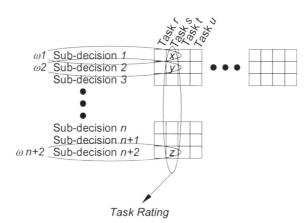

Task Rating

1. Take into account: specific changes in policy decisions, those policy areas and decisions
   predominantly considered strategically important to the business and the MS function,
   the possible policy area problems identified from the quick-hit table. These three factors
   help identify the important policy areas and decisions for the MSD project. Suitable
   weighting can then be applied.
2. Calculate the aggregate rating for each design task.
3. Prioritize the design tasks according to their rated values.

The rating for each task, with respect to the sub-decisions, is the sum of the product of the importance weightings for each particular sub-decision and its task value. If a strategic sub-decision is not considered, then its importance rating will be zero (see *Tool/Technique 2.4.1*). Hence, the ranking of a task with respect to the importance of strategic issues is as illustrated in the figure. The greater the relative value of this parameter with respect to the other design tasks, the greater the priority of the particular design task.

Similar approaches can be applied for the action plans and terms of reference tables. With respect to the action plans, these can simply be prioritized and given importance ratings by the user. The terms-of-reference is again similar in this respect:

1. Assign an importance weighting factor to the input criteria selected. The user is encouraged to resist the temptation to grade each criteria equally.
2. Calculate the aggregate weighting for each design task.
3. Prioritize the design tasks according to their task rating.

Since the MS strategy represents the definition of the role and operations of the MS function and its resources and supporting infrastructure, the MSA/MSD processes essentially define a future state of the systems and the means by which this will be achieved. However, due to the dynamic nature of competition, markets and business, the future intended state is likely to change and should therefore be under continuous review. The figure below illustrates these concepts, together with the principle of undertaking a number of MSD projects as part of the strategy implementation. In order to avoid local optimization and the threat of losing sight of the strategy and the end-goal, the strategic sub-decisions need to be considered during the specification and execution of each individual MSD project, whether strategic or tactical in nature.

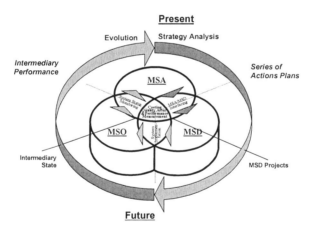

Finally, the sequencing of design tasks is very much dependent on the flow of data and composition of design decisions throughout the MSD process. The result of data analysis often determines the next step. However, placing the identified MSD tasks according to their task frames, as given in *Worksheet MSA/MSD 2.3.1*, normally provides a sensible plan of tasks within an MSD project.

TOOL/TECHNIQUE MSA/MSD 2.3.1—Action Plan/MSD Task Link

MSD Tasks	New systems for existing product(know-how existin	Modernisation	Innovation	Automation	Cost reduction	Rationalisation	Capacity expansions	Bottleneck elimination	Location change	Merging of sites	Decentralisation	Down sizing	Systems synchronisation	Systems modification	Diversification	New systems for new products	Joint ventures	Virtual factories	Simplify product line	Improve product design	Value analysis / product design	Design for manufacture	Reduce time-to-market for new products	Upgrade existing facilities	Improve equipment and process technology	Increase mechanisation	Increase capacity	Optimise make versus buy mix	Reduce materials losses	New processes for new products	New processes for old products	Recondition existing plants	Just-in-time manufacturing	Robotics and material handling	Introduce flexible manufacturign systems	Increase operations standardisation	Relocate plant	Improve energy/utilities efficiency	Link to business strategy	Improve vendors quality	Improve distribution	Supplier lead-time reduction	Reduce provisioning time	Improve equipment utilisation	Reduce order-to-delivery time for existing products	Lead-time reduction	Activity based costing
Product analysis	x	x	x	x	x	x	x	x	x	x	x	x	x	x	x	x	x	x	x	x	x	x		x	x	x	x	x	x	x	x	x			x			x		x	x	x		x	x		x
Part analysis	x	x	x	x	x	x	x	x	x	x	x	x	x	x	x	x	x	x	x	x	x	x	x	x	x	x	x	x	x	x	x	x			x			x		x	x	x		x	x		x
Process analysis	x	x	x	x	x	x	x	x	x	x	x	x	x	x	x	x	x	x	x	x	x	x	x	x	x	x	x	x	x	x	x	x			x			x		x	x	x		x	x		x
Functional make vs buy	x	x	x	x	x	x	x	x	x	x	x	x	x	x	x	x	x	x	x	x	x	x		x	x	x	x	x	x	x	x	x			x			x		x	x	x		x	x		x
Functional grouping	x	x	x	x	x	x	x	x	x	x	x	x	x	x	x	x	x		x	x	x			x			x		x	x		x			x	x		x		x		x		x	x	x	x
Capacity - demand	x	x	x	x	x	x	x	x	x	x	x	x	x	x	x	x			x	x				x	x		x	x		x	x		x			x			x		x	x	x		x	x	
Structural layout	x	x	x	x	x	x	x	x	x	x	x	x	x	x	x	x			x	x				x		x	x	x		x	x		x			x			x		x	x	x		x	x	
Integration - modularisation	x	x				x	x	x	x	x	x	x	x	x	x	x			x	x				x		x			x			x			x			x		x	x	x		x	x	x
Information functions	x				x	x		x		x	x				x	x			x	x			x			x		x			x				x			x		x	x	x					
Decision variables	x				x	x		x		x	x				x	x			x	x			x			x		x			x				x			x		x	x	x					
Process planning	x	x	x	x	x	x	x	x	x	x	x	x	x	x		x	x		x	x			x	x	x	x	x	x		x			x		x	x			x	x		x	x		x	x	x
Part grouping	x	x	x	x	x	x	x	x	x	x	x	x	x	x		x	x		x	x			x	x	x	x	x	x		x			x		x	x			x	x		x	x		x	x	
Make vs Buy	x	x	x	x	x	x	x	x	x	x	x	x	x	x		x	x		x	x			x	x	x	x	x	x		x			x		x	x		x	x	x		x	x		x	x	
Cell formation	x	x	x	x	x	x	x	x	x	x	x	x	x	x		x			x				x	x			x			x			x		x	x			x	x			x		x	x	
Conceptual layout	x	x	x	x	x	x	x	x	x	x	x	x	x	x		x			x				x	x	x	x			x			x		x	x			x	x			x		x	x	
Conceptual capacity	x	x	x	x	x	x	x	x	x	x	x	x	x	x		x			x				x	x	x	x			x			x		x	x			x	x		x			x	x	
Space determination	x	x	x	x	x	x	x	x	x	x	x	x	x	x		x			x				x		x	x			x			x			x			x	x					x	x	
Material Handling	x	x	x	x	x	x	x	x	x	x	x	x	x	x		x			x				x		x	x			x		x	x	x		x				x					x	x	
Factory storage	x	x	x	x	x	x	x	x	x	x	x	x	x	x		x			x				x		x				x		x	x			x				x					x	x	
Support	x	x	x	x	x	x	x	x	x	x	x	x	x	x		x			x	x			x	x	x	x	x	x		x			x		x	x	x			x	x	x	x	x	x	x	x
Facilities	x	x	x	x	x	x	x	x	x	x	x	x	x	x		x			x	x			x	x	x	x	x	x		x			x		x	x	x			x	x	x		x	x	x	
Integration	x	x					x			x	x	x	x			x				x	x			x	x			x			x		x	x			x		x		x		x	x	x	x
Autonomy	x	x					x			x		x	x			x				x				x				x		x		x			x			x		x		x	x				
Automation	x	x	x	x								x	x			x				x	x	x			x			x			x		x			x			x		x		x	x			
Planning and control	x	x	x	x	x	x	x	x	x	x	x	x	x	x		x			x				x			x			x		x	x	x		x	x	x			x	x		x	x	x	x	x
System architecture	x	x	x			x	x		x	x	x			x			x			x				x			x			x		x	x			x			x		x		x	x	x	x	x
Data flows	x					x	x		x	x	x			x			x			x				x			x			x		x			x			x		x		x	x			
Organisation structure	x		x				x	x	x	x			x	x			x			x				x			x			x			x			x			x		x		x	x	x	x	x
Organisation culture	x						x	x	x	x			x	x			x			x				x			x			x			x			x			x		x		x	x	x	x	x
Organisation state	x						x	x		x			x	x			x			x				x			x			x			x			x			x		x		x	x	x	x	x
Labour policy	x		x	x			x	x	x			x	x	x			x			x				x		x				x			x			x			x		x		x	x	x	x	x
Quality policy	x	x					x			x			x	x			x			x				x		x			x	x			x			x			x		x		x	x	x	x	x
Detailed domain layout	x	x	x	x	x	x	x	x	x	x	x	x	x	x		x			x				x	x			x	x		x	x		x	x	x	x	x	x			x					x	x
Detailed cell layout	x	x	x	x	x	x	x	x	x	x	x	x	x	x		x			x				x	x			x	x		x	x		x	x	x	x	x	x			x					x	x
Workstation layout	x	x	x	x	x		x	x	x	x	x	x	x	x		x			x				x	x	x	x			x	x		x	x	x	x	x	x			x					x	x
Equipment selection	x	x	x	x			x	x	x	x	x	x	x	x		x			x				x	x	x	x			x	x	x	x	x	x	x	x	x		x		x					x	x
Buffer sizes	x	x	x	x	x	x	x	x	x	x	x	x	x	x		x			x				x	x			x	x		x	x		x	x	x	x	x	x			x					x	x
Storage location	x	x	x	x	x	x	x	x	x	x	x	x	x	x		x			x				x	x	x	x			x	x		x	x		x	x	x			x					x	x
Storage system	x	x	x	x	x	x	x	x	x	x	x	x	x	x		x			x				x	x	x	x			x	x		x	x		x	x	x			x					x	x
Handling path	x	x	x			x	x		x	x	x	x	x	x		x			x				x	x			x	x		x	x		x	x	x	x	x			x					x	x
Handling unit	x	x	x			x	x	x	x	x	x	x	x	x		x			x				x	x			x	x		x	x		x	x	x	x	x			x					x	x
Maintenance	x	x	x	x	x	x	x			x		x	x	x		x			x				x	x			x	x		x	x	x	x	x	x	x	x	x						x	x	x	x
Tooling	x	x	x	x	x	x	x	x		x	x	x	x	x		x			x				x	x			x	x		x	x		x	x	x	x	x							x	x	x	x
Supplies	x	x	x	x	x	x	x			x	x	x	x	x		x			x				x	x			x	x		x	x	x	x	x	x	x	x	x			x	x	x	x	x	x	x
Administration	x	x					x	x	x			x	x			x			x				x	x			x	x		x	x		x	x	x	x	x			x	x	x	x	x	x	x
Set-up management	x	x	x	x	x			x		x	x	x	x	x		x			x				x	x			x	x		x	x		x	x	x	x	x							x	x	x	x
Process inspection	x	x	x	x	x			x		x	x	x	x	x		x			x				x	x			x	x		x	x		x	x	x	x	x			x	x	x		x	x		

(Continued on following page)

MSD Tasks	Production-inventory control systems	Decentralise decision making authority	Give workers planning tasks	Improve information handling	Interfunctional information systems	Improve MIS, finncl and oprtng systms, cntrls & reports	Manufacturing information systems	Introduce computer-aided technologies	Quality function deployment	Statistical process control	Statistical quality control	Quality circles	Zero defects	Establish total quality control programme	Reduce set-up times	Preventative maintenance	Reduce lost work time	Manufacturing re-organisation	Develop product workshops	New product introduction improvement	Implement group work	Interfunctional work teams	Improve communications	Improve departmental performance	Change organisational design/focus	Improve integration among departments/functions	Redesign jobs	Improve work methods and procedures	Ergonomics	Employee task enlargement/enrichment and responsibili	Develop a workforce with multiple, flexible skills	Improve mgr/supr/employee slctn, training and dvlpmnt	Worker / management / supervisor training	New skills hiring	Encourage employee involvement	Institute employee involvement with prdctvty gains-shari	Productivity bargaining	Improve union-mat relations & labour-related productivit	New wage system	Apply rewards and penalties	Work-environment improvement	Worker safety improvement	External environment improvement
Product analysis									x	x			x		x			x	x	x								x															
Part analysis									x	x			x		x			x	x	x								x															
Process analysis									x	x			x		x			x	x	x								x															
Functional make vs buy									x									x		x								x															
Functional grouping		x																x	x					x				x															
Capacity - demand		x																x	x					x				x															
Structural layout				x		x	x											x	x					x				x															
Integration - modularisation		x	x	x		x	x	x										x	x	x	x	x		x	x			x															
Information functions	x	x	x	x	x	x	x	x		x								x	x	x	x	x	x	x	x	x		x															
Decision variables	x	x	x	x	x	x	x	x		x								x	x	x	x	x	x	x	x	x		x															
Process planning																		x	x									x													x		x
Part grouping																		x	x									x															
Make vs Buy																		x	x									x															
Cell formation																		x	x		x							x															
Conceptual layout																		x	x		x							x														x	
Conceptual capacity																		x	x									x															
Space determination																		x	x									x	x												x	x	
Material Handling																		x	x									x	x													x	
Factory storage																		x										x															
Support		x	x					x										x	x																								
Facilities										x	x							x	x																						x		
Integration	x	x	x	x	x	x	x	x		x								x	x	x	x		x	x	x	x	x		x														
Autonomy	x	x	x	x	x	x	x	x		x								x	x		x		x	x	x	x	x	x		x	x												
Automation	x	x	x	x	x	x	x	x		x								x	x				x	x	x	x	x	x		x													
Planning and control	x	x	x	x	x	x	x	x		x	x							x	x			x	x	x				x															
System architecture	x	x	x	x	x	x	x	x		x	x						x	x	x	x		x	x	x				x															
Data flows	x	x	x	x	x	x	x	x		x	x						x	x		x		x	x	x				x															
Organisation structure		x	x		x	x	x		x		x		x	x		x		x	x	x	x	x	x	x	x	x	x	x		x	x	x	x	x	x	x			x	x			
Organisation culture		x	x		x		x		x	x	x	x	x		x		x	x	x	x	x	x	x	x	x	x	x	x		x	x	x	x	x	x	x	x	x	x	x	x	x	x
Organisation state		x	x		x		x		x		x	x	x		x		x	x	x	x	x	x	x	x	x	x	x		x	x	x	x	x	x	x	x	x	x	x	x	x	x	x
Labour policy		x	x							x	x	x	x		x	x		x	x		x	x	x	x		x	x	x	x	x	x	x	x	x	x		x	x		x			
Quality policy		x	x							x	x	x	x	x		x		x		x	x	x	x	x		x	x	x	x	x				x	x			x	x				x
Detailed domain layout																		x									x														x	x	
Detailed cell layout										x								x		x							x	x	x												x	x	
Workstation layout										x			x					x		x							x	x	x	x	x										x	x	
Equipment selection										x			x		x	x		x									x	x	x	x										x	x	x
Buffer sizes		x																x										x															
Storage location																		x									x	x														x	
Storage system													x					x									x	x														x	
Handling path					x												x	x									x	x														x	
Handling unit					x								x				x	x									x	x														x	
Maintenance		x	x					x			x		x	x	x	x		x		x							x	x	x	x	x				x								
Tooling		x	x								x		x	x	x	x		x		x							x	x	x	x	x												
Supplies		x	x					x			x		x	x	x	x		x		x							x	x		x	x												
Administration		x	x	x		x		x			x	x	x	x	x	x		x		x		x					x	x		x	x			x					x	x		x	
Set-up management		x	x								x		x		x			x		x							x	x	x	x	x								x	x			

TOOL/TECHNIQUE MSA/MSD 2.3.2—Strategic Decision/MSD Task Link

Policy Area	Decisions and Sub-decisions	Product analysis	Part analysis	Process analysis	Functional make vs buy	Functional grouping	Capacity - demand	Structural layout	Integration - modularisation	Information functions	Decision variables
	Total capacity										
C	Demand pitch					x				x	x
A	Floor Space					x					
P	Plant			x	x	x					
A	Equipment			x	x	x					
C	Labour					x					
I	**Variation Satisfaction**										
T	Cyclical	x	x	x	x	x					
Y	Long Term Trends				x	x					
	Demand Highs				x	x					
	Demand Lows				x	x					
	Degree of flexibility	x	x	x	x	x	x	x	x		
	Expansion Methods										
	How			x	x	x					
	Size of increment			x	x	x					
	Contraction Methods										
	How			x	x	x					
	Size of decrement			x	x	x					
	Timing										
	Bottlenecks										
	Demand forecasting										
	How monitor									x	x
	How forecast									x	x
	Cap. change signal									x	x
	Number					x					
F	**Specification**										
A	Size					x					
C	Capability	x	x	x	x						
I	**Location**										
L	Factory										
I	Facilities					x					
T	Plant Layout					x					
I	**Focus / specialisation**										
E	Type	x	x	x		x	x	x	x		
S	Degree	x	x	x		x	x	x	x		
	Function Integration										
	Enterprise						x	x			
	Manufacturing						x	x			
	Support services						x	x			
	Material Flow						x				
	Information flow									x	x

Policy Area	Decisions and Sub-decisions	Process planning	Part grouping	Make vs Buy	Cell formation	Conceptual layout	Conceptual capacity	Space determination	Material Handling	Factory storage	Support	Facilities
	Total capacity											
C	Demand pitch					x						
A	Floor Space					x	x	x	x			
P	Plant			x		x				x		
A	Equipment			x		x				x		
C	Labour					x						
I	**Variation Satisfaction**											
T	Cyclical			x						x		
Y	Long Term Trends											
	Demand Highs			x								
	Demand Lows			x						x		
	Degree of flexibility		x	x	x	x	x	x		x		
	Expansion Methods			x		x						
	How			x		x						
	Size of increment			x		x						
	Contraction Methods											
	How			x		x						
	Size of decrement			x		x						
	Timing											
	Bottlenecks											
	Demand forecasting											
	How monitor											
	How forecast											
	Cap. change signal											
	Number											
F	**Specification**											
A	Size											
C	Capability											
I	**Location**											
L	Factory											
I	Facilities											
T	Plant Layout					x	x					
I	**Focus / specialisation**											
E	Type		x	x	x	x						
S	Degree		x	x	x	x						
	Function Integration											
	Enterprise											x
	Manufacturing											x
	Support services									x		x
	Material Flow										x	x
	Information flow										x	x

Policy Area	Decisions and Sub-decisions	Integration	Autonomy	Automation	Planning and control	System architecture	Data flows	Organisation structure	Organisation culture	Organisation state	Labour policy	Quality policy
	Total capacity											
C	Demand pitch				x							
A	Floor Space											
P	Plant											
A	Equipment											
C	Labour							x		x		
I	**Variation Satisfaction**											
T	Cyclical							x		x		
Y	Long Term Trends							x		x		
	Demand Highs											
	Demand Lows											
	Degree of flexibility											
	Expansion Methods											
	How							x			x	
	Size of increment							x				
	Contraction Methods											
	How							x			x	
	Size of decrement							x				
	Timing											
	Bottlenecks											
	Demand forecasting											
	How monitor											
	How forecast											
	Cap. change signal											
	Number											
F	**Specification**											
A	Size							x				
C	Capability											
I	**Location**											
L	Factory											
I	Facilities											
T	Plant Layout											
I	**Focus / specialisation**											
E	Type											
S	Degree											
	Function Integration											
	Enterprise	x				x		x		x		
	Manufacturing	x				x		x		x		
	Support services	x				x		x		x		
	Material Flow	x	x	x	x							
	Information flow	x	x	x	x	x	x					

Policy Area	Decisions and Sub-decisions	Detailed domain layout	Detailed cell layout	Workstation layout	Equipment selection	Buffer sizes	Storage location Storage system	Handling path	Handling unit	Maintenance	Tooling	Supplies	Administration	Set-up management	Process inspection
	Total capacity														
C	Demand pitch														
A	Floor Space	x	x	x											
P	Plant				x										
A	Equipment				x										
C	Labour														
I	**Variation Satisfaction**														
T	Cyclical	x													
Y	Long Term Trends														
	Demand Highs														
	Demand Lows	x													
	Degree of flexibility														
	Expansion Methods														
	How	x	x	x	x										
	Size of increment	x	x	x	x										
	Contraction Methods														
	How	x	x	x	x										
	Size of decrement	x	x	x	x										
	Timing														
	Bottlenecks														
	Demand forecasting														
	How monitor														
	How forecast														
	Cap. change signal														
	Number														
F	**Specification**														
A	Size														
C	Capability														
I	**Location**														
L	Factory														
I	Facilities														
T	Plant Layout	x	x	x											
I	**Focus / specialisation**														
E	Type														
S	Degree														
	Function Integration														
	Enterprise	x		x											
	Manufacturing	x		x											
	Support services	x		x							x	x	x	x	x
	Material Flow	x	x	x					x	x					
	Information flow														

Policy Area	Decisions and Sub-decisions	Production planning	Scheduling	Software definition	Equipment selection	Batch sizes	Volume mixes	Shift patterns	Control systems	Data collection	Materials management	Software definition	Equipment selection	Job requirements	Job design	Training	Quality	Working conditions	Safety	Motivation	Reward systems	Human services	Material services	Machine services	Building	Health and safety
	Total capacity																									
C	Demand pitch	x	x																							
A	Floor Space																									x
P	Plant																						x		x	
A	Equipment																						x		x	
C	Labour							x										x	x			x				
I	**Variation Satisfaction**																									
T	Cyclical	x	x					x										x								
Y	Long Term Trends							x										x								
	Demand Highs							x										x								
	Demand Lows	x						x										x								
	Degree of flexibility																									
	Expansion Methods							x										x	x							
	How							x										x	x							
	Size of increment							x										x	x							
	Contraction Methods																									
	How							x										x	x							
	Size of decrement							x										x	x							
	Timing																									
	Bottlenecks																									
	Demand forecasting																									
	How monitor																									
	How forecast																									
	Cap. change signal																									
	Number																									
F	**Specification**																									
A	Size																									
C	Capability																									
I	**Location**																									
L	Factory																									
I	Facilities																									
T	Plant Layout																									
I	**Focus / specialisation**																									
E	Type																									
S	Degree																									
	Function Integration																									
	Enterprise																									
	Manufacturing																									
	Support services																									
	Material Flow																									
	Information flow	x	x	x																						

This is a very wide matrix chart with many columns. The column positions are labeled 1..N from left to right based on horizontal position of each mark.

Category	Row label	1	2	3	4	5	6	7	8	9	10	11	12	13	14	15	16	17	18	19	20	21	22	23	24	25
	Competitive type																									
S	**Time span**																									
U	**Sourcing type**																									
P	**Supplier**																									
P	Supplier qualification																									
L	Performance measurement																									
I	Supplier controls																									
E	Supplier selection																									
R	Supplier partnership					x					x															
S	Supplier assistance																									
	Technological cooperation																									
	Integration					x					x	x	x													
	Communication																									
	Products																									
	Components	x	x	x	x																					
	Services	x	x	x	x					x						x	x							x	x	x
H	**Culture - Behaviour**									x	x													x	x	
U	Time horizon									x	x															
M	Supervision							x		x	x	x							x	x				x	x	
A	Interdependence						x	x		x	x	x												x	x	
N	Risk taking attitude									x	x	x	x							x	x	x		x	x	
	Ownership of process							x		x	x	x	x	x	x	x	x	x	x	x				x	x	
R	Ownership of product							x		x	x	x	x	x										x	x	
E	Responsibility							x		x	x		x											x	x	
S	Comfort						x	x		x	x		x						x	x	x	x		x	x	
O	Teams						x	x		x	x								x	x		x				x
U	Communication				x		x	x		x	x	x							x	x	x			x	x	
R	**Production related**						x	x		x	x	x							x	x				x		
C	Quality concern									x	x	x	x	x						x	x	x	x	x	x	x
E	Process concern									x	x	x	x	x						x	x			x	x	
S	Productivity concern									x	x	x	x							x	x			x	x	
	Flexibility and change									x	x	x	x	x	x	x				x	x	x		x	x	
	Job content								x			x	x	x	x	x	x	x	x	x		x	x		x	x
	Cycle time											x	x	x	x					x	x		x	x		x
	Pacing											x	x	x						x	x		x	x		x
	Skill levels											x	x	x	x	x	x	x	x	x	x	x		x	x	
	Training									x		x	x						x	x	x					
	Motivation									x		x								x		x		x	x	
	General																									
	Employment security									x	x	x	x												x	
	Overtime policy									x	x	x	x												x	x
	Employee selection policy									x	x	x								x	x	x				
	Recruitment policy									x	x	x														
	Number of shifts			x				x		x		x							x		x	x				
	Safety									x		x							x	x			x	x		x
	Health											x										x	x			x
	Remuneration																									
	Payment systems									x	x	x	x											x	x	
	Payment structures									x	x	x	x	x										x	x	
	Pay ranges											x							x						x	
	Incentives and rewards									x	x	x	x							x				x	x	x
	Implications																									
	fulfilling manufacturing role																									
	alternatives																									

Code	Row label	1	2	3	4	5	6	7	8	9	10	11	12	13	14	15	16	17	18	19	20
Q	Implementation - Extent						x														
U	Product quality																				
A	Design																				
L	Design process																				
I	Process quality																				
T	Capability vs inspection																	x			
Y	Inspection locations																	x			
	inspection frequency																				
	Training																	x			
	Monitoring																				
	Total quality																				
	Initiatives																				
	Documentation																				
	Training																				
	Responsiblity																		x		
	Quality levels																				
	how select																				
	levels selected																				
	Implications																				
	fulfilling manufacturing role																				
	alternatives																				
	Supplier relations - Inventory		x		x					x		x	x		x				x		
P	Inventory																				
L	Size			x						x									x		
A	Spread	x	x	x					x										x		
N	Balance	x	x	x					x										x		
N	Location			x					x			x							x		
I	Function		x	x				x		x		x		x			x		x		
N	Manufacturing priorities																				
G	Method					x	x														

Code	Row label	continued																		
	Organisational level					x	x													
	Centralisation					x	x													
&	Co-ordination					x	x													
	Autonomy					x	x													
	Response level					x	x													
C	Management																			
O	Materials management					x	x													
N	Customer promises					x	x													
T	Forecasting																			
R	System					x	x													
O	Investment					x														
L	Planning - Time horizon					x	x													
	Scheduling																			
	Time horizon					x	x													
	Resource allocation					x	x													
	Formal paradigms					x	x													
	Informal					x	x													
	Centralisation					x	x													
	Monitoring					x	x													
	Updating time frame					x	x													
	Control					x	x													
	Order release					x	x													
	Expediting					x	x													
	Batch sizes					x	x													
	Production approach			x		x	x						x	x		x			x		

Policy Area	*Decisions and Sub-decisions*	Product analysis	Part analysis	Process analysis	Functional make vs buy	Functional grouping	Capacity - demand	Structural layout	Integration - modularisation	Information functions	Decision variables	Process planning	Part grouping	Make vs Buy	Cell formation	Conceptual layout	Conceptual capacity	Space determination	Material Handling	Factory storage	Support	Facilities	Integration	Autonomy	Automation	Planning and control	System architecture	Data flows	Organisation structure	Organisation culture	Organisation state	Labour policy	Quality policy	Detailed domain layout	Detailed cell layout	Workstation layout	Equipment selection	Buffer sizes	Storage location Storage system	Handling path	Handling unit	Maintenance	Tooling	Supplies	Administration	Set-up management	Process inspection	Production planning	Scheduling	Software definition	Equipment selection	Batch sizes	Volume mixes	Shift patterns	Control systems	Data collection	Materials management	Software definition	Equipment selection	Job requirements	Job design	Training	Quality	Working conditions	Safety	Motivation	Reward systems	Human services	Material services	Machine services	Building	Health and safety
P	**Product details** - Scope	x	x	x	x							x	x	x																																																										
R	Focus	x	x	x	x							x	x	x																																																										
O	Range	x	x	x	x							x	x	x																																																										
D	Volume	x	x	x	x							x	x	x																																																										
U	**Introduction**																																																																								
C	Introduction rate					x		x	x						x	x							x	x	x	x	x		x	x	x			x	x	x	x						x	x	x			x							x		x			x	x											
T	Introduction philosophy	x	x	x						x	x																		x	x	x		x																																							
S	Product life cycle length					x		x	x						x	x							x	x	x	x	x							x	x	x	x						x	x	x			x							x					x	x											
	Computer aids								x	x	x												x				x	x																																												
	C-A application								x	x	x												x				x	x		x	x		x				x										x								x	x		x	x				x									
	C-A extent								x	x	x												x				x	x	x	x	x																																									
	Innovation								x																				x	x	x																																									
	Lead-times																																																																								
	Product design																												x	x	x																																									
	Manufacturing	x	x	x	x	x		x	x			x	x	x	x	x			x				x						x	x	x									x			x	x	x	x	x					x	x							x	x											
P	**Performance** - Criteria									x	x																x	x	x	x	x		x																							x																
E	Compet. variables focus									x	x																x	x	x	x	x		x																							x																
R	Business mgt intg focus									x	x												x				x	x	x	x	x																									x																
F	Benchmarking									x	x																x	x	x	x	x		x																							x										x	x					
O	**Structure** - General					x		x	x																				x	x	x																																									
R	Flatness																												x			x																																								
G	Formality																												x	x	x																																									
A	Centralisation																												x																																											
N	Control																												x																																											
I	**State**																														x	x	x																													x	x			x	x					
S	**Management**																																																																								
A	Openness																													x																																										
T	Product understanding																																																																								
I	Manufacturing understanding																																																																								
O	Systems perspective																												x	x	x																																									
N	Culture																													x																																										
	Functions																																																																								
	Emphasis					x																							x																																											
	Management supervision					x																		x					x																																											
	Co-ordination																																																																								
	Marketing					x			x	x																			x																																															
	Engineering					x			x	x																			x																																															
	Customers					x			x	x																			x																																															

TOOL/TECH. MSA/MSD 2.3.3—Terms of Reference/MSD Task Link

Option	Product analysis	Part analysis	Process analysis	Functional make vs buy	Functional grouping	Capacity - demand	Structural layout	Integration - modularisation	Information functions	Decision variables	Process planning	Part grouping	Make vs Buy	Cell formation	Conceptual layout	Conceptual capacity	Space determination	Material Handling	Factory storage	Support	Facilities	Integration	Autonomy	Automation	Planning and control	System architecture	Data flows	Organisation structure	Organisation culture	Organisation state	Labour policy	Quality policy	Detailed domain layout	Detailed cell layout	Workstation layout	Equipment selection
SCOPE																																				
Product / system																																				
EP-ES	x	x	x	x	x	x	x	x			x	x	x	x	x	x	x	x	x	x	x	x	x	x	x	x							x	x	x	x
ENP-ES	x	x	x	x	x	x	x	x			x	x	x	x	x	x	x	x	x	x	x	x	x	x	x	x							x	x	x	x
NP-ES	x	x	x	x	x	x	x	x			x	x	x	x	x	x	x	x	x	x	x	x	x	x	x								x	x	x	x
EP-NS	x	x	x	x	x	x	x	x	x	x	x	x	x	x	x	x	x	x	x	x	x	x	x	x	x	x	x	x	x	x	x	x	x	x	x	x
ENP-NS	x	x	x	x	x	x	x	x	x	x	x	x	x	x	x	x	x	x	x	x	x	x	x	x	x	x	x	x	x	x	x	x	x	x	x	x
NP-NS	x	x	x	x	x	x	x	x	x	x	x	x	x	x	x	x	x	x	x	x	x	x	x	x	x	x	x	x	x	x	x	x	x	x	x	x
Project type																																				
Redesign	x	x	x	x	x	x	x	x	x	x	x	x	x	x	x	x	x	x	x	x	x	x	x	x	x	x	x	x	x	x	x	x	x	x	x	x
Optimisation	x	x	x	x	x	x	x	x	x	x	x	x	x	x	x	x	x	x	x	x	x	x	x	x	x	x	x	x	x	x	x	x	x	x	x	x
New system	x	x	x	x	x	x	x	x	x	x	x	x	x	x	x	x	x	x	x	x	x	x	x	x	x	x	x	x	x	x	x	x	x	x	x	x
Step improvement	x	x	x	x	x	x	x	x	x	x	x	x	x	x	x	x	x	x	x	x	x	x	x	x	x	x	x	x	x	x	x	x	x	x	x	x
Continuous improvement	x	x	x		x	x	x	x			x	x	x																				x	x	x	
New philosophy																																				
Project focus																																				
Physical	x	x	x	x	x	x	x	x			x	x	x	x	x	x	x	x															x	x	x	x
Control									x	x												x	x	x	x	x	x	x								x
Decisional									x	x												x	x	x	x	x	x			x						
Support																x	x	x									x	x	x	x						
Organisational					x					x																		x	x	x	x					
System integration										x								x		x		x	x	x	x	x	x	x	x	x			x	x	x	
Solution drivers																																				
Cells	x	x	x	x	x	x	x	x			x	x	x	x	x	x	x	x	x	x	x	x	x	x	x	x	x	x	x	x	x	x	x	x	x	x
CIM	x	x	x	x	x	x	x	x	x	x	x	x	x	x	x	x	x	x	x	x	x	x	x	x	x	x	x	x	x	x	x	x	x	x	x	x
Group technology	x	x	x	x	x	x	x	x			x	x	x	x	x	x	x	x	x	x		x	x	x	x	x	x						x	x	x	x
Just In Time	x	x	x	x	x	x	x	x	x	x	x	x	x	x	x	x	x	x	x	x	x	x	x	x	x	x		x	x	x	x		x	x	x	x
Total Quality Management																																x	x	x	x	x
OBJECTIVES																																				
Cost Related																																				
reduce production cost	x	x	x	x																																
reduce overheads				x									x			x	x	x	x	x													x	x	x	x
reduce direct to indirect labour ratio																								x	x			x		x			x	x	x	x
Operations Related																																				
reduce lead-time	x	x	x	x	x	x	x	x			x	x	x	x	x	x	x	x	x	x	x			x	x	x		x	x	x			x	x	x	x
improve schedule adherence	x	x	x						x	x														x	x	x		x	x	x			x	x	x	x
increase throughput	x	x	x	x	x	x	x	x			x	x	x	x	x	x	x	x	x	x													x	x	x	x
increase productivity																																	x	x	x	x
reduce inventory level	x	x	x	x	x	x	x	x			x	x	x	x	x	x	x	x							x									x	x	
reduce work in progress	x	x	x	x	x	x	x	x			x	x	x	x	x	x	x								x									x	x	
reduce non value adding activities																					x							x	x	x	x					
increase level of automation	x	x	x	x	x	x	x	x	x	x	x	x	x	x	x	x	x	x				x	x	x	x	x	x						x	x	x	x
increase process flexibility	x	x	x	x	x	x	x	x			x	x	x	x	x	x	x	x															x	x	x	x
increase volume flexibility	x	x	x	x	x	x					x	x	x	x	x																		x	x	x	x
increase production volume	x	x	x	x	x	x	x				x	x	x	x	x	x																	x	x	x	x
increase level of integration	x	x	x	x	x	x	x															x	x	x	x	x	x	x	x				x	x	x	x
Orgnl. Restructure																																				
simplify material flow	x	x	x	x	x	x	x	x			x	x	x	x	x	x	x	x	x	x	x	x											x	x	x	x
rationalise product range	x	x	x	x	x	x					x	x	x																				x	x	x	x
increase manufacturing focus	x	x	x	x	x	x	x				x	x	x	x		x	x					x						x					x			
increase stdisation & commonality	x	x	x	x	x		x	x			x																									
increase level of decentralisation					x			x	x	x												x	x		x	x	x	x	x	x			x	x	x	x
simplify control system								x	x	x						x						x	x	x	x	x	x						x	x	x	
improve ownership of product	x	x	x	x	x			x	x		x	x		x														x	x	x	x	x				
Quality Related																																				
improve quality																																	x	x	x	x
reduce waste	x	x	x	x																													x	x	x	x
achieve zero defects																																	x	x	x	x

Option	Buffer sizes	Storage location	Storage system	Handling path	Handling unit	Maintenance	Tooling	Supplies	Administration	Set-up management	Process inspection	Production planning	Scheduling	Software definition	Equipment selection	Batch sizes	Volume mixes	Shift patterns	Control systems	Data collection	Materials management	Software definition	Equipment selection	Job requirements	Job design	Training	Quality	Working conditions	Safety	Motivation	Reward systems	Human services	Material services	Machine services	Building	Health and safety
SCOPE																																				
Product / system																																				
EP-ES	x	x	x	x	x	x	x	x	x	x	x	x	x			x	x	x	x	x	x			x	x	x	x	x		x	x	x	x		x	x
ENP-ES	x	x	x	x	x	x	x	x	x	x	x	x	x			x	x	x	x		x			x	x	x	x			x	x		x	x		x
NP-ES	x	x	x	x	x	x	x	x	x	x	x	x	x			x	x	x	x		x			x	x	x	x			x	x		x	x		x
EP-NS	x	x	x	x	x	x	x	x	x	x	x	x	x	x	x	x	x	x	x	x	x	x	x	x	x	x	x	x	x	x	x	x	x	x	x	x
ENP-NS	x	x	x	x	x	x	x	x	x	x	x	x	x	x	x	x	x	x	x	x	x	x	x	x	x	x	x	x	x	x	x	x	x	x	x	x
NP-NS	x	x	x	x	x	x	x	x	x	x	x	x	x	x	x	x	x	x	x	x	x	x	x	x	x	x	x	x	x	x	x	x	x	x	x	x
Project type																																				
Redesign	x	x	x	x	x	x	x	x	x	x	x	x	x	x	x	x	x	x	x	x	x	x	x	x	x	x	x	x	x	x	x	x	x	x	x	x
Optimisation	x	x	x	x	x	x	x	x	x	x	x	x	x	x	x	x	x	x	x	x	x	x	x	x	x	x	x	x	x	x	x	x	x	x	x	x
New system	x	x	x	x	x	x	x	x	x	x	x	x	x	x	x	x	x	x	x	x	x	x	x	x	x	x	x	x	x	x	x	x	x	x	x	x
Step improvement	x	x	x	x	x	x	x	x	x	x	x	x	x	x	x	x	x	x	x	x	x	x	x	x	x	x	x	x	x	x	x	x	x	x	x	x
Continuous improvement	x			x	x	x	x	x	x	x	x								x	x				x	x	x	x			x	x		x	x		
New philosophy																																				
Project focus																																				
Physical	x	x	x	x	x																x			x	x			x	x			x	x	x	x	
Control												x	x						x	x	x	x	x	x	x	x					x		x	x	x	x
Decisional												x	x	x	x	x	x	x						x	x					x	x					
Support						x	x	x	x	x	x										x			x	x			x	x							
Organisational																			x					x	x	x	x	x	x	x	x	x				
System integration	x	x	x	x	x	x	x	x	x	x	x	x	x	x	x	x	x	x	x	x	x	x	x	x	x					x	x	x	x	x	x	x
Solution drivers																																				
Cells	x	x	x	x	x	x	x	x	x	x	x	x	x	x	x	x	x	x	x	x	x	x	x	x	x	x	x	x	x	x	x		x	x	x	x
CIM	x	x	x	x	x	x	x	x	x	x	x	x	x	x	x	x	x	x	x	x	x	x	x	x	x	x	x	x	x	x	x	x	x	x	x	x
Group technology						x	x	x	x	x	x	x	x	x	x	x	x	x	x	x	x	x	x	x	x	x	x	x	x	x	x		x	x	x	x
Just In Time	x	x	x	x	x	x	x	x	x	x	x	x	x	x	x	x	x	x	x	x	x	x	x	x	x	x	x	x	x	x	x	x	x	x	x	x
Total Quality Management						x	x	x	x	x	x								x	x	x			x	x	x	x	x	x	x	x	x	x	x	x	x
OBJECTIVES																																				
Cost Related																																				
reduce production cost																																				
reduce overheads						x	x	x	x	x	x													x	x		x	x	x		x	x	x	x	x	x
reduce direct to indirect labour ratio						x	x	x	x	x	x	x	x	x	x				x	x	x	x	x	x	x	x	x		x	x	x	x	x			
Operations Related																																				
reduce lead-time	x	x	x	x	x	x	x	x	x	x	x	x	x	x	x	x	x	x	x	x	x	x	x	x	x	x	x	x	x	x	x	x	x	x		
improve schedule adherence				x	x	x	x	x	x	x	x	x	x	x	x	x	x	x	x	x	x			x	x	x	x			x	x	x	x			
increase throughput																																				
increase productivity						x	x	x	x	x	x	x	x						x	x	x							x	x	x	x	x	x			
reduce inventory level	x	x	x	x	x		x	x	x	x	x	x	x	x	x	x	x	x	x	x	x	x	x													
reduce work in progress	x	x	x	x	x				x	x	x	x	x	x	x	x	x	x	x	x	x	x	x													
reduce non value adding activities				x	x	x	x	x	x	x	x								x	x	x			x	x	x	x					x	x	x		
increase level of automation	x	x	x	x	x	x	x	x	x	x	x	x	x	x	x	x	x	x	x	x	x	x	x	x	x	x	x	x	x			x	x	x	x	x
increase process flexibility	x					x	x	x	x	x	x					x	x	x						x	x	x	x					x	x			
increase volume flexibility	x					x	x	x	x	x	x					x	x	x						x	x	x	x					x	x			
increase production volume	x	x	x	x	x	x	x	x	x	x	x					x	x	x						x	x		x					x	x			
increase level of integration	x	x	x	x	x	x	x	x	x	x	x	x	x	x	x	x	x	x	x	x	x	x	x	x	x	x	x									
Orgnl. Restructure																																				
simplify material flow		x		x			x	x				x	x			x	x		x			x			x	x						x	x	x		
rationalise product range																																				
increase manufacturing focus									x			x	x																			x	x	x		
increase stdisation & commonality																																				
increase level of decentralisation						x	x	x	x	x	x	x	x	x	x				x	x	x	x	x	x	x					x						
simplify control system												x	x	x	x	x	x		x	x	x	x	x	x	x	x	x		x							
improve ownership of product																								x	x		x		x							
Quality Related																																				
improve quality				x	x	x	x	x	x	x	x					x	x	x						x	x	x	x	x		x	x		x	x		
reduce waste		x		x	x	x	x	x	x	x	x															x	x			x			x	x		
achieve zero defects					x	x	x	x	x	x	x													x	x	x	x						x	x		

WORKSHEET MSA/MSD 2.3.1—MSD Task Selection

Project Title:

Person(s) Responsible:

Version: **Date Completed:**

MSD project title:

	MSD Task		Task Description
REQUIREMENTS	*System Function*	1. 2. 3.	
	System Structure	1. 2. 3.	
	System Decisions	1. 2. 3.	
CONCEPTUAL DESIGN	*Manufacturing & Supply Processes*	1. 2. 3.	
	Human & Organization	1. 2. 3.	
	Information & Control	1. 2. 3.	
DETAILED DESIGN	*Processes*	1. 2. 3.	
	Facilities	1. 2. 3.	
	Supports	1. 2. 3.	
	Planning	1. 2. 3.	
	Control	1. 2. 3.	
	Human	1. 2. 3.	
	Organization	1. 2. 3.	
	Warehouse & Transportation	1. 2. 3.	

WORKSHEET MSA/MSD 2.3.2—MSD Task Objectives

Project title: **MSD Task:**

Person(s) Responsible:

Version: **Date Completed:**

MSD Stage

Requirement specification	
Conceptual design	
Detailed design	

MSD Task Focus

Physical systems		Support systems	
Control systems		Organizational systems	
Decisional systems		Systems integration	

Task Objectives

Cost Related

Reduce production cost		Reduce materials costs	
Reduce direct/ indirect costs		Other.........	
Reduce overheads			

Operations Related

Reduce lead-times		Increase production volume	
Increase productivity		Increase throughput	
Reduce NVA activity		Reduce work in progress	
Increase Volume flexibility		Increase process flexibility	
Improve schedule adherence		Increase integration	
Reduce inventory		Other.........	
Increase automation			

Organization Related			
Simplify flow		Simplify control system	
Rationalize product range		Improve ownership	
Standardize/ commonality		Increase decentralization	
Other ….			

Quality Related	
Improve quality	
Reduce waste	
Achieve zero defects	
Other………	

Task Constraint

Time Constraints

Resource Constraints

Human Resource Constraints

Financial Constraints

Task Objectives

Objectives	Desirability	Feasibility	Availability	Priority	Targets

Task Document MSA/MSD 2.4—Planning of Project Execution

TASK OVERVIEW

TASK DESCRIPTION

Based on the outlines previously specified, this produces detailed plans for the execution of the project(s) from start to finish, taking into consideration all the MSD and the subsequent MS implementation activities. A breakdown of the project tasks may be necessary so that the components can be assigned to different individuals or groups for completion. Also, requirements on human resources have to be identified, and training needs specified. To define the work content of a project, the activities that have to be performed need to be listed at a sufficient level of detail. The dependencies among them are then determined, the critical path is determined, the resources are allocated, and the project gateways are defined. In all, the task consists of the following steps:

Identify project activities. Using work breakdown (WBD) and organizational breakdown (OBD) diagrams to define a breakdown of activities and responsibilities.

Determine activity dependencies. Determining the relations and links between the identified activities.

Determine critical path and gateways. Establishing the critical path. This makes it easier to schedule activities and allocate resources. To be beneficial, it should be continuously updated throughout the course of project execution, so as to give management the chance to reallocate resources to the activities which are falling behind schedule, and to those which may become the new critical path. In relation to this, gateways represent the termination of important activities. The purpose of the gateways is to provide feedback about the project. Normally, gateways can be set according milestones such as: the agreement of project plan, the completion of a conceptual design, the acceptance of a detailed design, the completion of operational procedures, the completion of a factory building, the installation of equipment, the agreement of organizational changes, and the completion of training.

In reality, project management software packages (such as Microsoft Project) are available to plan and manage a project. The use of these tools is highly recommended. The output from this stage can be used directly as the input to such software packages.

TASK LINKS

POSITION IN MSM FRAMEWORK

INPUT FROM :	OUTPUT TO :
MSA/MSD 2.1 (project list)	MSD execution.
MSA/MSD 2.2 (term-of-ref.)	MS implementation.
MSA/MSD 2.3 (project tasks)	

OUTPUTS

Detailed project plans for design execution.

Overall plan of system implementation.

MSA

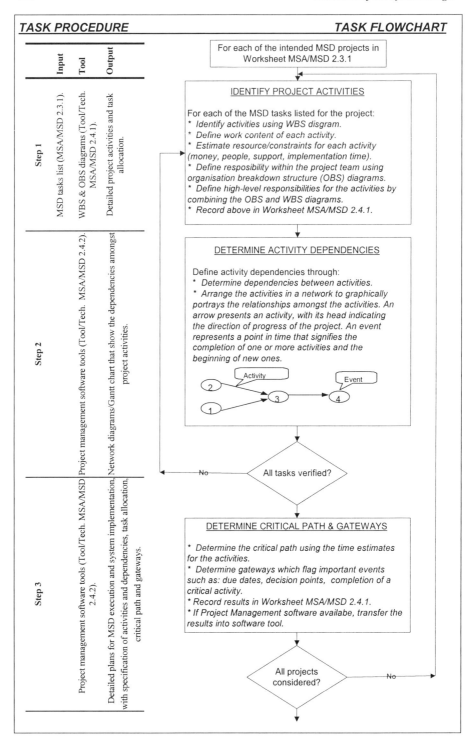

TOOL/TECHNIQUE MSA/MSD 2.4.1—WB and OB structures

Since real-life projects tend to become complex, it may prove difficult to coordinate individuals working on different parts, and to keep track of all of the components. This is particularly true if the project concerns a distributed MS system where the facilities are positioned at a number of different locations. Work breakdown structure (WBS) is a tool for defining a hierarchical breakdown of work contents and responsibilities in an MSD project. It is developed by: (1) *Breaking down all the high level tasks of a project into more detailed*

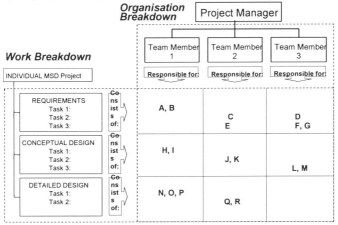

PROJECT ACTIVITIES

levels. The tasks include the previously identified MSD tasks plus the MS implementation actions needed. The end units of such WBS are known as project activities; (2) *Estimating the work required to complete the activities.* This can be assigned to a project team member/unit. Organization breakdown structure (OBS) can then be used to define task and responsibility allocation for the project. In general, an OBS model shows a project's organizational structure. It defines the communication lines used for reporting progress from the bottom up, and the lines for issuing work orders and technical instructions from the top down. Such a structure can be assembled by specifying the different levels of organization responsible for the execution and implementation of the project. The higher levels must represent various management layers, from project team leaders and department managers up to top-level management. The lowest levels represent operational units engaged in the execution of activities. The OBS and WBS can then be combined to define authority and responsibility. Together they specify which departments/units should be involved with which project activities, so that each project activity is linked to both structures at their lowest levels as shown in the figure. Procedures for work authorization, report preparation and distribution are thus established at the appropriate level. The procedures that can be followed to establish this structure include:

- Identify availability of resources at the three MS layers: manufacturing/supply processes, information flows, and human resources.
- Allocate resources amongst projects/activities.
- Verify whether the time plan is feasible with respect to available resources.
- Determine if resources outside the company (new personnel, equipment) are needed to accelerate design and implementation activities.

- • Determine if training is required for employees involved with the new system.
- • Identify training needs and determine relevant training programs.
- • Reshape and redefine roles, behaviors and responsibilities in the new organization.

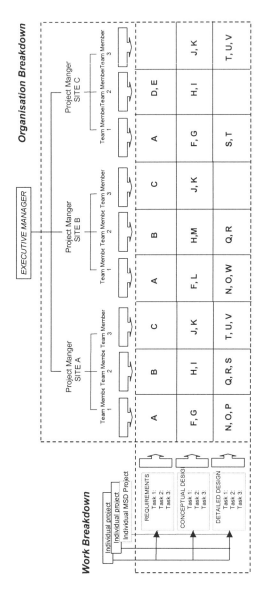

For initiatives of distributed MS systems involving a multitude of MSD projects and different sites, a similar WBD/OBD matrix may also be employed for the purpose of project planning, as shown above.

WORKSHEET MSA/MSD 2.4.1—Project Execution Plan

MSD Project Title: **MSD Task:**

Person(s) Responsible:

Version: **Date Completed:**

Project Work Breakdown Structure

	MSD and Implementation Tasks in Project	Activity Specification			
		I D	Activity Description	Wk Load (Days)	Resource & Constraints
REQUIRENEMTS	System Function MSD Tasks				
	System Structure MSD Tasks				
	System Decisions MSD Tasks				
CONCEPTUAL DESIGN	Manuf'g and Supply Processes MSD Tasks				
	Human and Organization MSD Tasks				
	Information and Control MSD Tasks				
DETAILED DESIGN	Processes				
	Facilities				
	Supports				
	Planning				
	Control				
	Human				
	Organization				
	Warehouse and Transport				
IMPLEMENTATION	Manuf'g and Supply Processes				
	Human and Organization				
	Information and Control				

Project Organization Breakdown Structure

		Task Group or Dept. (Name):			Task Group or Dept. (Name):			Task Group or Dept. (Name):			Task Group or Dept. (Name):		
Project Leader/Owner (Name):														
Site: **Location:**														
Organization Breakdown Structure / Work Breakdown Structure		Member (Name):	Member (Name):	Member (Name):	Member (Name):	Member (Name):	Member (Name):	Member (Name):	Member (Name):	Member (Name):	Member (Name):	Member (Name):	Member (Name):
	Project Activity ID												
REQUIRENMTS	System Function												
	System Structure												
	System Decisions												
CNCPTUL DSGN	Manufacturing and Supply Processes												
	Human and Organization												
	Information and Control												
DETAILED DESIGN	Processes												
	Facilities												
	Supports												
	Planning												
	Control												
	Human												
	Organization												
	Warehouse and Transport												
IMPLEMENTATION	Manufacturing and Supply Processes												
	Human and Organization												
	Information and Control												

CHAPTER FOUR

Execution of MS System Design Tasks

4.1 INTRODUCTION

Progressing through the previous MSA/MSD processes will have helped to answer questions like: *where we are now? where do we need to be? and which route do we take?* Now it is time to tackle the key issue of *how to get there*, through the execution of the previously chosen MSD project(s). This chapter discusses how to find the best structure for an MS system: one which will support the strategic objectives under the constraints specified. Firstly, this chapter outlines the principles involved in the execution of MSD tasks and their respective outputs within the overall MSM reference architecture of system design. A generic MSD task document (*Task Document MSD 1*) will be provided as a template to help the execution of various MSD tasks. Secondly, the general techniques of the tasks in each of the main design areas are presented, and, where appropriate, worksheets and checklist provided. Together, these provide a complete set of tools to help the execution of the necessary MSD tasks as identified in the required project(s).

Specifically, this chapter outlines the principles of the MSD tasks involved in each of the six design areas, and provides generic worksheets to aid in their execution. In comparison with the previous task documents, the task documents presented here will provide only a generic template. In practice, the user may need to tailor it to suit the specific MSD tasks required by their project. However, specific worksheets are provided in each of the areas. Detailed accounts of the individual MSD tasks, including specific analysis techniques and tools, have been presented previously in *Manufacturing and Supply Systems Management* (Wu 2000). The reader is, of course, advised to consult other sources of information where necessary.

4.2 MSD PROBLEM-SOLVING CYCLE: A GENERIC TASK DOCUMENT

The general procedure for executing structured MSD tasks within the framework is shown in Figure 4.1. In general, design is fundamentally the process of creating, evaluating and selecting an alternative. Regardless of the problem to be addressed, the execution of an MSD task should follow a problem-solving cycle, as depicted in the figure.

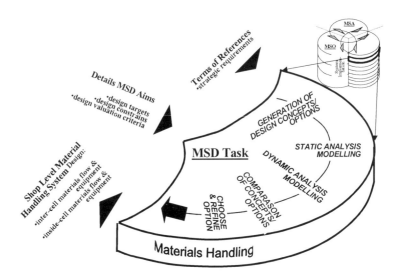

Figure 4.1 The problem-solving cycle of MSD task execution

As can be seen, the technique breaks the total task into a set of broad steps, and demands certain fixed outputs from one stage before logically continuing to the next:

- *Generation of design concepts* requires initiative to create a relatively comprehensive set of alternatives. The number of ideas produced should be as large as possible under the time and resources constraints. Initially, judgment at an intuitive level is sufficient for a first-pass analysis of these ideas to identify any candidates which appear to meet the strategic/task objectives, and at the same time, not to violate constraints. Following the above, the aim of the evaluation of concepts is to identify which solutions have the greatest outcome value—as measured by the performance criterion—for the least risk. This process involves the most scientific elements in the cycle of systems analysis. The tasks involved here can be divided into two categories: model building and outcome evaluation.
- *Model building* is needed to provide the analytical tools. The type of modeling techniques used to evaluate the alternatives is diverse, and includes mathematical, physical and simulation models. However, their application here can be divided into two main groups: for *static analysis*—to evaluate the design options' capabilities of satisfying the general demands upon the system; and for *dynamic analysis*—to predict the options' transient behavior and, hence, the ability to cope with the dynamic operating conditions.
- *Outcome evaluation*. With the help of a properly constructed model, the performance of the system under each of the alternatives may be tested via a comparison of either quantitative or qualitative results. For any MS system design under consideration, there are two sets of criteria to be assessed. The first question is to ask whether the system fulfills the requirements initially specified through the MSA processes. The second criterion in any real setting will be that

of financial justification—whether the system will generate enough returns to justify the investment. On the basis of the evaluation, it is possible to make a rational decision about whether to implement the system, consider further development of the design, or terminate the task.

The generic MSD task document (*Task Document MSD 1*) and its worksheets follow this structured problem-solving approach. It can be adopted to help the execution of the majority of MSD tasks within the MSM framework. Although the following discussion focuses on the design of manufacturing processes, the principles and techniques are usually equally applicable to the supply aspects of an organization.

4.3 MSD TASK OUTPUTS WITHIN THE MS SYSTEM STRUCTURE

Typically, the engineering design of a product requires a number of documents to be produced: part drawings to define the geometrical features of the items such as shape and dimension, part lists and drawings to show how these parts should be assembled, and procedures to specify how the product should be tested and operated. The requirement is identical when specifying an MS system. The complete specification has to include a number of documents, each providing design information about a specific part of the system—in this case, in the areas of physical facilities, information, human resources and organization. In addition, rules and cross-checking mechanisms should be provided to show how different parts of the system should be interrelated to guarantee system integrity, and how they should function cooperatively when put into practice.

As outlined in Chapter 1, the complete MSM framework consists of two domains: the MSM tasks specifying the analytical and design processes, and the reference architecture providing the logical basis for the complete specification of a manufacturing and supply system. The three related phases of the overall structure shown in Figure 1.8 represent the main design steps: *system requirement definition, conceptual design,* and *detailed design.* The first defines the system boundary, the second develops the basic principles by which the system will work, and the third provides detailed accounts of what is required and, hence, a complete design. The outputs, which are the results from various tasks along the MSA/MSD cycle, are summarized in Figure 4.2. As shown, the results from the relevant MSA and MPM tasks will have specified the overall strategic requirements for the system, together with detailed targets for the MSD tasks (the core area). The execution of the MSD tasks then provide the detailed contents for system design.

With a greenfield project, one starts with the set of objectives and then creates a system model that fits the intended purpose with little need to consider an existing system. More often, however, when the projects concerned are of the brownfield or continuous improvement types, one has to consider the existing system, analyze its structure, and then try to modify it to fulfill the future requirements. This allows the incorporation of experience already gained, but the options available could be constrained and the ideas limited. In either case, through the route-planning of the previous MSA/MSD process, one or more MSD projects will have been specified. Each of these projects contains a number of related design tasks relevant for a particular stage in the design process, as well as a particular MS architecture or sub-architecture. With the help of the reference model of MS system structure shown in Figure 4.2, the user can identify the elements of results from each of the relevant MSD tasks and validate their relationships within the complete structure of an MS organization.

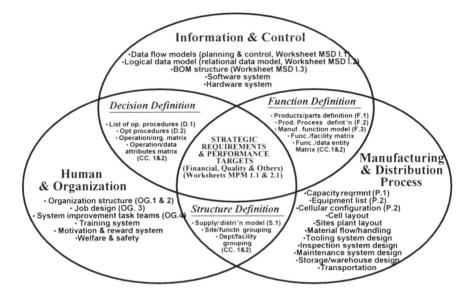

Figure 4.2 Overall reference structure for the complete specification of an MS system

Generally speaking, the MSD tasks at the system requirement definition stage are associated with the conceptual design of an MS system, producing results and decisions that outline the overall purpose, characteristics and structure of the system. The results from these tasks include system models that specify required manufacturing and logistic functions. Each of these functions has a related catalog of products, together with a hierarchy of control systems that process information. In addition, the conceptual modeling specifies the long-term production capacity to be achieved in terms of the average or static capacity levels:

- The *function* specification outlines what to make versus buy, and how to make or buy it, as required. It defines the system's boundary of operation and draws a map of the transformation and supply processes through products and parts definition, definition of capacity requirement, manufacturing and supply function model (e.g., IDEF$_0$ or SCOR—a supply chain specification—model).
- The *structure* specification outlines the overall system structure in terms of system sites and their geographical location and layout.
- The *decision* specification provides the operational procedures required to run the system.

In relation to the above, the MSD tasks at the subsequent conceptual/detailed design stages will specify, list and organize the system entities at the three system layers. The detailed design stage essentially transforms the conceptual model into detailed specifications. To summarize, there are three main areas to be considered in the detailed design stage: the selection of production and supply technology, together with the selection of transportation and storage facilities; the organization and layout of the technology; and the detailed design of the control system, including both hardware and software. The output from this stage will be a design which is accurate to a high level, and detailed enough for the actual system implementation. The results from this stage includes:

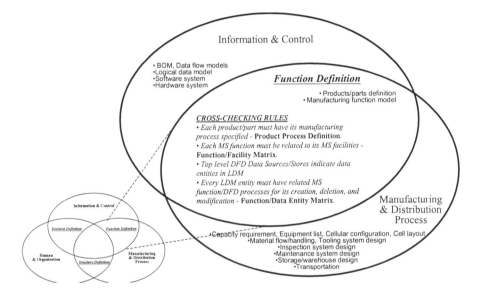

Figure 4.3 Function definition—cross-checking between MS functions and information

- The physical elements of the ***manufacturing and supply process*** area: equipment list, cellular configuration, cell layout, material flow definition, material handling process, tooling system design, inspection system design, maintenance system design, storage and warehouse design, and transportation, etc. These are items utilized by the system to carry out the transformation and supply processes.

- The manufacturing and supply information system at the ***information and control*** layer, whose structure and contents can be specified using the standard methods such as *data flow diagram (DFD)* to specify its functionality, and the *entity-relationship (ER)* model to define its database structure. In addition, the *software* and *hardware* need to be chosen or developed for the system's implementation.

- The ***human and organization structure*** layer describes the structure of the entity, including: *organization structure* (in terms of systems sites, departments and personnel), *job design, training procedures*, and other human resource policies and practice, as shown.

In reality, the completion of a system design is unlikely to be achieved sequentially, since MSD decisions in each area will have implications for the others. Therefore, some of the results produced within the three overlapping layers also serve the purpose of defining the nature of the interactions between the two layers involved, and thus provide a logical means of system-wide cross-checking. Through a number of iterations, the validity of each of the individual layers, as well as the overall integrity of the entire system structure, can be guaranteed:

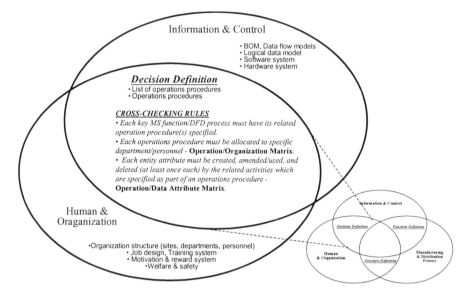

Figure 4.4 Decision definition—cross-checking between MS organization and information

- The ***function*** definition provides cross-checking between products/parts and facilities through *production processes* and *function/facility matrix*; and between functions and the information system through *function/data entity matrix* (Figure 4.3)
- The ***decision*** definition specifies the interaction between the information and the organization structures through the *organization/operation matrix*, and the *operation/data entity matrix*. The former defines the roles and responsibilities of the employees, in terms of the cross-relationship between the organization (departments and/or personnel) and the operational functions within the system. The later specifies the relationship between the operational procedures and the data entities of the information system. Together, these two matrices define the employees' responsibility and access for data operation, decisions and MS functions (Figure 4.4).
- The ***structure*** definition further specifies the organizational structure and responsibility by mapping the cross-relationship between the organizational departments and the MS processes: the *site/function matrix* helps to clarify which MS functions are to be located on which site; and the *department/facility matrix* specifies which MS equipment and facilities are required by which departments (Figure 4.5).

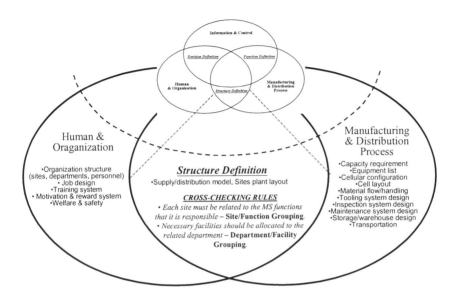

Figure 4.5 Structure definition—cross-checking between MS organization and functions

4.4 MSD TASKS—SYSTEM FUNCTIONS

The MSD tasks of this group aim to produce a conceptual design for the MS system under consideration. Such a model covers the system in a general sense and develops the basic principles by which the systems will operate. It does this by specifying the activities necessary for the system to perform its intended task. It thereby provides a framework of further decomposition by outlining the basic building blocks. These blocks will be comprised of a combination of the required manufacturing and supply functions and, to a certain extent, the necessary controlling functions.

4.4.1 General Process

The main process involved here is shown in Figure 4.6. For illustration purposes, only a few MSD tasks are outlined in this diagram. There a few MSD tasks belonging to this area that are not shown on the diagram. However, they should also follow this general path to enhance the design.

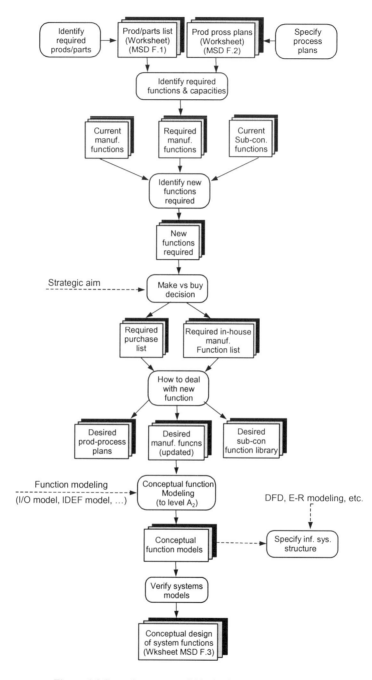

Figure 4.6 General processes within the *function* MSD domain

Engineering analysis of products (output: products/parts definition)

The market requirements will have defined the product range and the competitive stance, and these will have a major influence upon the system to be designed. The desired product range may include new products, enhanced products, and different quantities of current products. The information gathered about these should include the products' expected parts lists. Each component part should be identified and recorded in a desired part catalog. Estimates of demands for all products should also be obtained. A current product catalog should be created. This should identify the quantities of finished parts to be dispatched, showing cyclical variations if necessary. For each product listed in the catalog, a part list should be produced, allowing creation of a current part catalog specifying every part that must be manufactured or procured, together with the demand levels, as illustrated in Figure 4.6.

Process analysis (output: production process definition)

By considering the manufacturing process for each part identified, it is possible to identify the manufacturing functions needed in the system. The necessary functions should be recorded in a desired manufacturing function library. Next it is necessary to identify the different types and capacities of functions required in the manufacturing system. Identification of process plans for each part in the current part catalog allows the functions to be identified. These functions should be recorded as a manufacturing function list. When cross referenced to the current part catalog, it is then possible to calculate the total capacity demands for each function using estimated operation duration. It will also be possible to cross reference the manufacturing functions to the plant register, and thus give a reflection on the ability of the system to provide for the required processes.

Analysis capacity requirement (output: capacity requirement)

By comparing the currently available manufacturing capacity (both in-house and subcontracted) to the desired manufacturing capacity, the currently unavailable manufacturing functions/capacities can be identified. Before further action can be taken, it is necessary to decide how the expertise to support these functions will be provided. There are three options available: bringing it in from an outside company, developing the expertise in-house and subcontracting the work.

Manufacturing function modeling (output: manufacturing function model)

This consists of the *physical systems description* and the *control systems description*. Following the above, the manufacturing functions at this stage may be modeled and described using input/output cascade (or the $IDEF_0$ technique—see Section 4.4.2). This allows the process plans to accurately represent flows from one department or function to another. The departments identified will need further decomposition later to allow full assessment of the problems. The control functions can then be described following the information flows. The company's current operating procedures will provide the starting point for this analysis.

The results from the above will provide a functional specification of the system being analyzed and designed. The model should be in greater detail in areas that are expected to require further analysis and design actions. In general, the resultant functional specification of the system from the above should be checked against the structure of the prototype system model and the associated prerequisite conditions described in Section 4.4.3. This will help highlight areas that may be inconsistent, and therefore, likely to be sources of problems. Specifically, the strategic issues previously specified for the system should also be taken into consideration to guide its construction, and to make sure it will fulfill the requirements.

4.4.2 Function Modeling Tool: $IDEF_0$

$IDEF_0$ is a tool that can be used for the functional specification of an MS system. An IDEF model is a structured representation of the functions of the system and the flow of material and information which interrelate to these functions. The basic element of an $IDEF_0$ model is called a function block, such as the one shown in Figure 4.7.

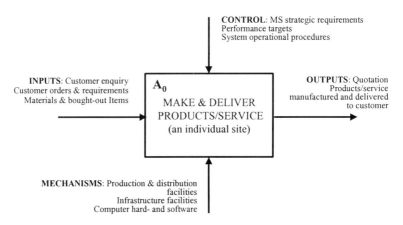

Figure 4.7 A top-level function block

The individual function blocks are linked together through the inputs, the outputs, the mechanism and the controls. When an input is utilized to create an output, a function will be actuated. The performance of the function is carried out through a mechanism and under the guidance of the control. The inputs to a function entering the function block from the left are usually (but not necessarily) consumed by the function to produce outputs. Raw materials are typical examples of these. The mechanism, represented by an arrow entering the function block from below, indicates the resources which are required to carry out the transformation process—such as machines, trucks, operators and drivers. All resources shown must be used as means to achieve the function. Finally, the controls which enter from the top of the block only influence the transformation process and will not be consumed or processed themselves.

Taking advantage of the hierarchical characteristics of an MS system, it is by nature a top-down approach. That is, it exposes one new level of detail at a time, beginning at the highest level by modeling the system as a whole. At the uppermost level, a function block is usually labeled as function A_0, which represents the overall system objectives and system boundary. In accordance with the hierarchical nature of a system, an $IDEF_0$ model can be decomposed level by level to describe each of the sub-systems within the structure, and this can be done to any level of detail. If, for example, A_0 consists of four sub-functions, then they will be called A_1, A_2, A_3, and A_4. Each of these sub-functions, together with their associated inputs, outputs, controls and resources, may themselves be decomposed into the next level in the hierarchy. The sub-function blocks at the next level will be named as A_{11}, A_{12}, ..., A_{21}, A_{22}, ..., and A_{41}, A_{42}, ..., etc. This provides a means of decomposing and allows a function of the system to be examined in detail while maintaining overall perspective. Thus, it allows the emergent properties of a system to be recognized at all times.

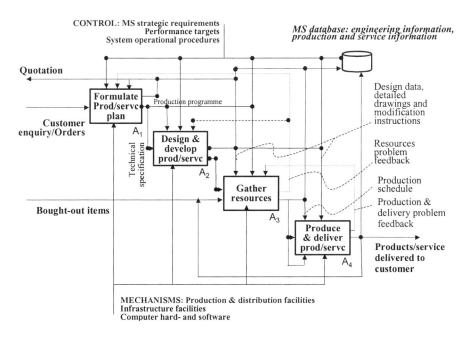

Figure 4.8 Level-1 decomposition of a function model

A description of a top-level function will identify the purpose of the system, and the competitive stance to be taken. The inputs will normally include all the materials and parts which are bought in for the MS process. These can then be organized to produce a more detailed system structure by decomposing the top-level model until the level concerning the component MS function is reached. The outputs will include a summary of the information given in the desired part catalog. For example, the top-level function model of Figure 4.7—which reveals the general context and structure of an MS organization characterized by make-to-order production and delivery services—may be decomposed into the following four areas (Figure 4.8):

- **Formulating production/service plan,** involving the sales department, costing control, design office, and production planning departments of an MS organization.
- **Designing and developing product/service to order**, involving the design office, development, testing and quality departments.
- **Gathering resources**, involving the purchasing and stock control department.
- **Producing and delivery products/service**, including parts producing activities, sub-assembly and final assembly operations, logistics and transportation. These involve the machine shops and assembly lines, the production control, and the dispatch departments.

From a systems perspective, therefore, the functional structure of this example MS system may be described from the uppermost level, A_0, down to its lower levels of decomposition, as listed below:

A_0 MAKE AND DELIVER TO CUSTOMER ORDER

A_1 FORMULATE MANUFACTURING/SERVICE PLAN

A_{11} Sale and Contract

A_{12} Plan Production Schedule

A_{13} Plan Delivery Schedule

A_2 DESIGN AND DEVELOP PRODUCT/SERVICE TO ORDER

A_{21} Control Design and Development Process

A_{22} Develop Prototype

A_{221} prepare advanced drawings

A_{222} make and test prototype(s)

A_{223} prepare final drawings and part lists

A_3 GATHER RESOURCES

A_{31} Plan Material and Capacity Requirements

A_{32} Gather Resources

A_{321} acquire production capacities

A_{322} acquire materials and bought-out items

A_4 PRODUCE AND DELIVER PRODUCTS/SERVICE

A_{41} Control Production Activities

A_{42} Carry Out Production Activities

A_{421} produce parts of products

A_{422} produce sub-assemblies of products

A_{423} produce final assembly

A_{424} test final assembly

A_{43} Deliver Products To Customer

A_{431} prepare and pack products

A_{432} transport and deliver products

4.4.3 Conditions for Effective System Structure and Operation

The conceptual design plays the most important role in determining the nature and characteristics of the system. Attention should be focused on the structure of the existing system, including its elements, relationships, boundaries, environment, and functions, as well as its strengths and weaknesses. Based on the structure of a generic system model, this section provides a list of pre-conditions for the effective systems operation. This list can be used to help check the soundness of a conceptual MS model under consideration.

Despite their diversity, all systems have some characteristics in common. This has led to the development of systems thinking: an attempt to explain the fundamental structure and nature of systems in a logical way. A key feature is the concept of viewing the situation or domain from a global perspective and of breaking this down into separate functions, at the same time taking their relationships into consideration. Fundamentally, the system analyst/designer should: (a) develop an understanding of a prototype system structure and keep a mental picture of this in mind; (b) whenever relevant, try to recognize and analyze a situation with such a system's perspective; (c) try to apply the structure and the associated pre-conditions of a prototype model to assist in the search for effective system solutions.

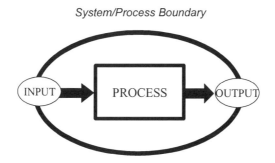

Figure 4.9 System viewed as a process/function

An MS system can be viewed as a collection of processes which are interrelated in an organized way and cooperate towards the accomplishment of the strategic ends. That is, it consists of a collection of transformation processes which convert a set of inputs to a set of outputs, as shown in Figure 4.9. The inputs and outputs are the main interfaces amongst the processes, and between the system itself and the outside world. The MS system is the totality of such processes and their relationships. A MS system is hierarchical in nature, because the system at one level can be a sub-system or even a component of higher systems. For instance, a number of systems can normally be identified within an MS company at a departmental level. It is apparent that all of the systems at this particular level must operate within the company system, which is one level up in the hierarchy and, hence, an upper system of the departmental systems. Conversely, depending on the number of hierarchical levels involved, a system at a particular level of the hierarchy may be further divided into sub-systems and components, each of which will receive inputs and transform them into outputs. For example, within a departmental system, each of its task teams may be considered as a sub-system. The relationship between a sub-system and the system is equivalent to that between the system and its upper system. That is, a sub-system can be a total system in itself, consisting of all the components, attributes and relationships necessary to achieve the objective which the upper system has mandated. The company itself can be a system within the upper system of a business corporation. An upper system influences its constituent system by laying out its operational goals, checking its performance and supporting its operation. In relation to this view, a checklist of pre-conditions for its effective structure and operation can be provided as follows.

The required overall system/sub-system structure

Manufacturing/supply systems are open systems. Such systems must have a set of operational processes which regulate or control the system's operational processes through communication of information. In system terms, this is the feedback-control function. System feedback takes place whenever information about any of the system's outputs is used to correct its operation. The essential components of a typical feedback control, within the MS context, include those illustrated in Figure 4.10.

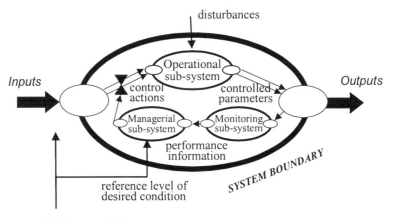

Figure 4.10 Overall structure of a functional system

These components include:

- an MS function that results in a controlled system parameter or condition,
- a monitoring function which measures the current status of the condition,
- a decision-making function that compares the current state of the condition with a desired goal/objective, and

- a control action that, when necessary, changes the MS operation towards the achievement of the desired goal.

Production control, for instance, is one of the many feedback controls exercised within an MS environment. Other examples include: quality control, cost control and purchase control. Therefore, for any system/sub-system to function, *all of the necessary parts as shown in Figure 4.10 must exist within the system boundary.* Also, feedback control may appear at more than one level. A higher level control governs the lower levels by monitoring their overall performances and setting the desired reference levels for them. The concept of such a control hierarchy is closely related to the concept of system hierarchy, as shown in Figure 4.11.

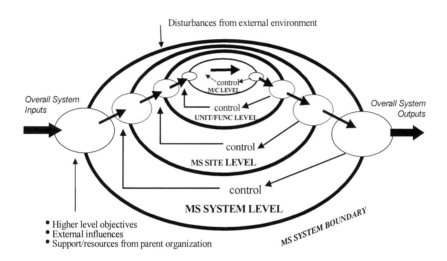

Figure 4.11 Hierarchy of control within an MS operation

Effective communication mechanisms

Effective communication is one of the prerequisites for successful control. Internally, the various sub-systems must communicate effectively in order to achieve successful control and policy/decision implementation. To increase overall system efficiency, close external links must be established between the organization and its customers, suppliers, and any other relevant bodies,. Accordingly, *effective communication mechanisms must be specified along this hierarchical structure of control.* Communications within an MS organization take place either vertically or horizontally. Vertical communication includes both downward and upward pathways of information flow, corresponding to the two portions of the control loop. Communications can also take place along horizontal paths at the department-to-department and person-to-person levels. These are mostly concerned with the process of actual input/output transformation. Communications within an MS organization can be established through either human-activity or physical-activity based means. The former are normally associated with higher levels of the management hierarchy, such as meetings and discussions, or communication through telephone, e-mail, or other e-business means. The later are associated with computer-based process control of machines and other facilities.

Adequate sub-system structures

The necessary sub-systems should be designed and implemented properly so that they perform their intended tasks adequately:

- *Adequate understanding of the MS transformation processes.* To design and implement a control mechanism, the process to be controlled must be understood to a required level of technical detail, including its inputs, outputs, flows, states, behavior, etc.
- *Ability to cope with disturbances.* Sufficient resources and flexible utilization should be employed for the key functions. As reflected by one of the basic rules of JIT philosophy, a focus on the provision of sufficient capacity, rather than its level of utilization, is necessary to cope with unpredictable disturbances from the market and environment.
- *Adequate measurement of the transformation processes.* According to the objectives or goals of the organization, one must be able to measure relevant process parameters in an adequate manner. This applies equally down the control hierarchy. That is, the strategies and policies adopted at various levels of the organization must be coherent, and the choice of measurement and the frequency and accuracy of measurement must be in line with the overall operational aims.
- *Appropriate managerial sub-system.* The managerial sub-systems must be capable of making the right decisions for the particular processes being controlled. In addition to human issues, clearly defined decision/operational procedures play an important role.

In fact, it should be apparent to the reader that the MSM framework closely follows the principles outlined above, as reflected by its processes for coherent strategy and goal-setting, by its structure of closed-loop MS management, and by the contents of its system reference model.

4.5 MSD TASKS—SYSTEM STRUCTURE

Having defined how the parts are to be made in terms of the required functions and capacity allocation (in-house or externally), the identified functions must next be organized in such a way that the objectives laid down can be fulfilled effectively. The general processes involved are as shown in Figure 4.12. The top layer model will present the system operation as a single function with identified inputs and outputs. A number of different groupings can be achieved at the lower levels of decomposition, dependent on the criteria applied to the decomposing process. Hence, a few system options may be generated with different organizational structure and technologies. All of these, however, aim to fulfill the same set of outputs, and each will have a different chance of success.

Decomposition by site—supply and distribution network modeling

As a major MS design decision, the make-or-buy decision should be considered for each of the major parts involved. Subcontracting, for example, has several advantages. It is a major method of increasing the flexibility of capacity; it can be used to provide extra capacity during peak periods or even meet 100% of the requirements for a particular function, thus allowing the company to develop and fully utilize its own expertise. It also allows the provision of manufacturing expertise which is outside the range of the current MS system so that a wider range of technologies may be utilized. The additional advantages include reduced inventory and reduced short-term risks. Potential problems include hidden costs—such as that of managing the infrastructure required—and hidden dangers, related to the lack of control over quality and delivery. Once the decisions are made, a subcontract function register should be created to formally record manufacturing functions which are available to the system, but outside the organization.

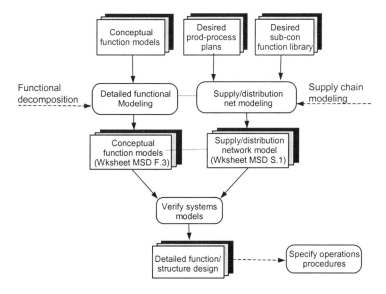

Figure 4.12 General processes within the *system structure* MSD domain

In relation to the make-or-buy decision are the problems of facility location planning which deals with the problems of geographical location of the production facilities, the location of distribution facilities, and supply management. A well planned distributed MS network allows a company to take advantage of economical, financial and technological factors related to facilities located in different geographical locations. The aim is generally to plan and coordinate all the MS activities necessary to provide the customers with required service levels at the minimum possible cost. This is done through coordination of information and material flow from the market place and from the suppliers to the manufacturing system, and from the manufacturing system back to suppliers and customers. A supply network potentially involves a number of manufacturing plants, warehouses, distribution depots, and the actual transportation between suppliers and customers. The following need to be taken into consideration:

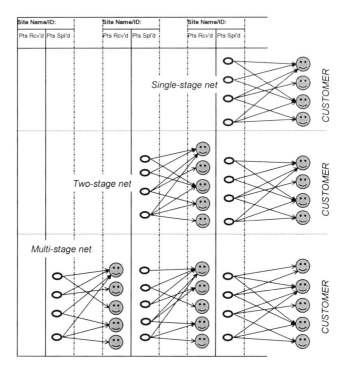

Figure 4.13 Different types of supply/distribution network

- **Types of distribution network.** When designing the structure of a supply network, issues such as the number of materials/parts/products to be handled, capacity restrictions, and the number of stages in the logistics network all have an impact on the final solutions. The structure of a distribution network itself can be specified according to: (a) a *single echelon network* that can involve either one- or two-stage networks; and (b) a *multi-echelon network,* as shown in Figure 4.13. The design of one-stage networks does not need to consider inbound transport. In contrast, the structuring of multi-echelon networks consists of various levels of facilities between a set of sources and a set of clients, dealing with the simultaneous location of manufacturing plants, warehouses and depots.

- **Manufacturing location analysis.** Manufacturing location analysis depends on plant location factors, which can be grouped or summarized under the following categories: (a) *Transfer costs,* which result from the movement of raw material and finished products to and from the plants to market; (b) *Production costs,* which include all expenses necessary to convert raw materials into finished goods, and which are usually variable and dependent on the geographical location; (c) *Maintenance costs,* which will again be different for different site locations; (d) *External economies of location*, which refer to cost reductions resulting from the geographical clustering of sites; (e) *Intangible location factors*, which include items such as personal contacts, influences of management, human needs and desires.

Decomposition by product/process matrix

Manufacturing technology has been pictured as a continuum ranging from process industry through production lines—large and small batch—and finally, to job shops (Figure 4.14). While this helps to identify the general technology requirement, a number of different technologies and approaches may be applied for the same product/process combination. Flexible manufacturing systems and group technology, for example, are two ideas which have been applied to the mid-volume/mid-variety part of this continuum.

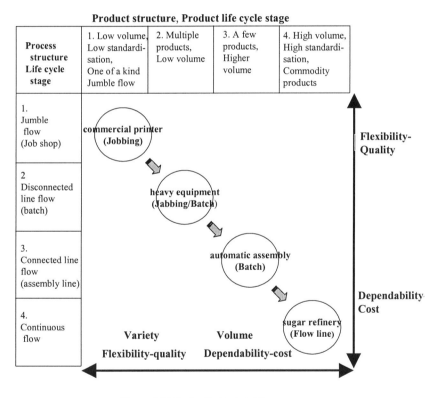

Figure 4.14 Product/process matrix

Decomposition by competitive characteristics

Certain parts of the product range may fall into different categories due to variation in the ways that products compete in their market places. Thus, some products require high quality and others require low cost, although the manufacturing functions are essentially the same. Therefore, the competitive characteristics of the products/Product groups, as identified at the MSA stages, may be used as the criteria for the decomposition of a system, so that necessary facilities can be offered according to particular requirements.

4.6 MSD TASKS—SYSTEM DECISION

Each physical function defined in the conceptual model will require that its own internal/external control system be elucidated. Different areas of the system will have different requirements for the type of control needed. Important considerations here include the level of synchronization required, the amount of information to be processed by the system, and the time required for processing. To achieve satisfactory operation, it is essential that the different control systems be effectively coordinated.

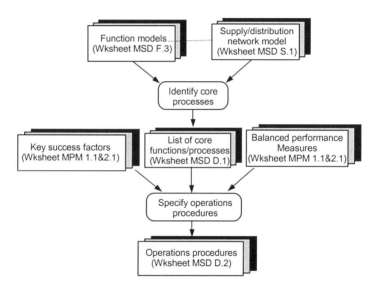

Figure 4.15 General processes within the *system decision* MSD domain

In general, an operational procedure should be specified using *Worksheet MSD D1* and *D2* for each of the core functions identified from the functional model (Figure 4.15). As can be seen, such operational procedures help to specify the activities to be carried out, the decision processes to be followed, the parameters to be controlled, and the targets to be achieved by the function concerned. As shown in Figure 4.16, based on the strategic requirements of the system, the MSA and the MPM processes will have specified specific measures and targets for their operations (*Worksheets MPM 1.1* and *2.1*). This stage of system design is where such objectives are transferred into operational criteria and tied to the various levels of systems management, thus becoming an integral part of system operation. Therefore, the performance measures specified in *Worksheet MSD 2.1* should always be tied to the system's current goals and objectives. Again, the representation of the required controlling functions on the physical system can be achieved in a top-down manner, with a control function superimposed on each level of the decomposition. These control functions can be further decomposed to provide a more detailed description of the information processing involved. The collection of completed operational procedures may be used for a number of purposes: e.g., as operational manuals to help decision-making and controlling of system functions, or as the basis for evaluating, implementing and operating a software system such as an ERP. Note that a control system need not always be computer based. The Kanban card method, for example, is a common non-computerized control approach.

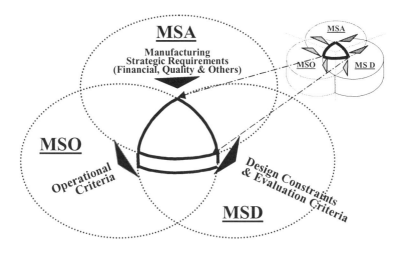

Figure 4.16 Operational criteria derived from MS strategy

In practice, operational procedures are traditionally paper-based. However, hyper-media technologies are increasingly being used for documentation and management within the intranet environment of an organization. The generic structure of a task-centered, multi-media information (TCMM) system for such purposes is shown in Figure 4.17. With a task-centered user interface, online referencing, digital manuals and an integrated computer-based training (CBT) module, a web-based documentation system can be used to provide a user-friendly information environment. Such a system can be used at various levels within an MS organization as a reference library to provide information about product data and operational procedures; a task-centered, interactive system to help carry out online operations; and a computer-aided training tool to train the company's managers/operators. Such technology provides facilities for the electronic format of documentation and its distribution, and allows the system to combine the capabilities of formerly separate entities such as animation, graphics, video, text, etc. With an open system structure, the system can link documents of various types in a task-centered way. More specifically, the main features of such a system are:

- ***Electronic documentation***. A digitized reference library provides information such as product data and operational procedures (Module a). A web-based database of reference manuals provides a means of supplying company personnel with comprehensive tools for looking up procedures and product information. A web-based document management system, delivered through the organization's intranet, also solves some of the problems associated with paper-based documents. An obvious advantage is the reduction of effort and cost in updating and maintaining the contents of the system. Once the electronic workbook of operational procedures is up and running, any site connected to the network can access the most relevant and up-to-date information. The same access may also be achieved through a CD-ROM based approach.
- ***Task-centered approach.*** The task-centered concept may be used to provide all the information relating to a particular function/process online directly at the point where the tasks are to be executed. This allows the user to navigate through the system as required and to access the relevant information in a focused way. The efficiency can be further enhanced by providing photographs or video of complex setups, special fixture configurations, etc.

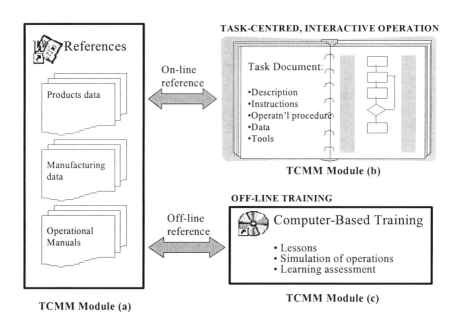

Figure 4.17 Generic structure of a TCMM system for the documentation
and management of MS operational procedures

Such an approach also provides a means of "institutionalizing" the MSM procedures within an MS organization. Cases of application in industry can be found in Chapter 7.

4.7 MSD TASKS—PHYSICAL FACILITIES

Some of the main tasks involved here include: *MS technology acquisition, selection of MS machines and facilities, cellular formation* and *cell/plant layout, material-handling, warehouse and transposition design.* The general processes involved are as shown in Figure 4.18.

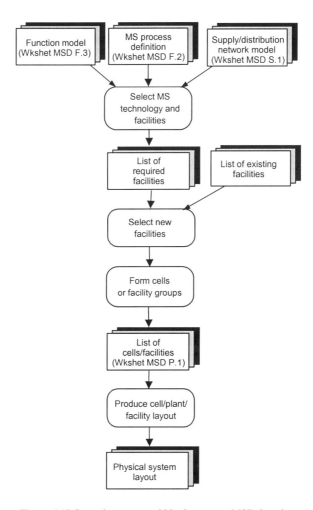

Figure 4.18 General processes within the *process* MSD domain

Selection of facilities

The requirements specified by the conceptual model provide guidance to appropriate technology. Several of the identified functions may be fulfilled by a single machine. Nevertheless, more than one machine is usually available to serve a particular MS function. It is therefore necessary to provide detailed specifications for the selection of most suitable items of the plant. Normally it is necessary to utilize the current plant to minimize the costs involved whenever possible. Each item of the plant should be considered for its possible application by assessing it against the hierarchy of criteria which have been identified previously.

Having allocated the current plant items, certain capacity requirements will remain unfulfilled or only partially fulfilled. Therefore, it is next necessary to establish what new MS technology and facilities should be used to satisfy the remaining requirements. It is first necessary to assess what equipment is currently available on the market. This is one of the best opportunities to look for innovative options, since there are less constraints attached. The results from the SWOT should be taken into consideration for the analysis.

Organization and layout of facilities

This consists of facility grouping and physical layout. The grouping of facilities is important, particularly their organization into cells. The aims and techniques of cellular formation can be found easily in the literature (e.g., Wu 1994). In general, the objectives of the physical layout of cells and other facilities should be in agreement with the overall objectives, and will often fall within one of three categories:

- Minimization of the cost of materials handling and movement,
- Minimization of congestion and delay, and
- Maximized utilization of space, facilities and labor.

The key here is *simplicity*. It is particularly important to simplify material flows when distributed-MS and advanced-MS systems are concerned. This must be achieved within cells/sites, as well as between them. To achieve the best layout of work-centers within cells, and the location of cells and departments in relation to one another, the space requirements of the previously identified functional groupings should first be established. These should be estimated on the basis of expected floor space for each of the plant items previously identified.

Following the above, the individual site and its departments must now be positioned. The decisions can be made using both the quantitative information generated and the constraints identified earlier. The decisions must be recorded as a map of the locations.

Warehouse location and transportation analysis

The number and geographic locations of warehouses are determined by manufacturing locations and markets, as specified by the *supply-distribution network model*. Warehouses can be classified as follows:

- *Market-positioned warehouses* are close to the market served in order to replenish inventory rapidly and at the lowest cost of transportation.
- *Production-positioned warehouses* are located close to manufacturing plants, so as to improve customer service.
- *Intermediately positioned warehouses* are located between customers and plants to achieve a balance between customer service and distribution cost.

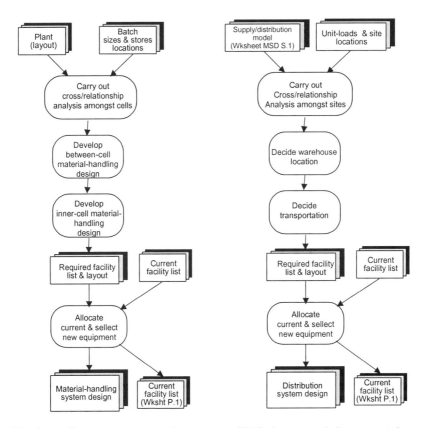

Figure 4.19 General MSD processes of distribution and material handling

Frequently, transportation and warehouse issues should be taken into consideration simultaneously (Figure 4.19). The main requirement or advantage of adding warehouse in a supply system is to reduce distribution cost and/or improve customer service level. So far as the transportation economies are concerned, the following general rules apply:

- *Warehouse justification.* A single warehouse is considered as a consolidation point for transportation shipment. A sufficient volume of shipments has to be available to justify the fixed cost of the warehouse facility.
- *Transportation cost minimization.* As the warehouse is added, total transportation cost decreases. As long as the total cost of warehousing, including local delivery, is equal to or less than the total cost of direct shipments to customers, the facility is economically viable.

Materials Handling

The concepts and techniques regarding materials handling are relevant within the boundary of the entire manufacturing and supply system. They can be used to analyze and design the materials handling system of a particular site, or employed to tackle the same problems across the entire supply chain. Materials handling may be defined as the techniques employed to move, transport, store, or distribute materials, with or without the aid of mechanical devices, with three main aspects:

- *Materials flow:* the flow of materials into, through, and away from the MS system.
- *Management:* the effective planning, control, review and improvement of the movements, handling and storage of materials.
- *Technology:* the techniques associated with the movement, handling and storage of materials. This MSD area also covers the issues related to the materials handling within cellular-based manufacturing environments.

The very complexity of the materials handling aspects of an MS system and their overbearing influence on the resulting systems performance demands that decisions taken in this area are closely related to organizational objectives if the performance of the organization is not to be impaired. The task should begin with a step to analyze the system's strategic requirements, and consists of a number of interrelated steps. Typically these include:

- system requirement analysis,
- material flow analysis,
- unit load selection,
- inter-cell equipment selection,
- inside-cell equipment selection, and
- system evaluation.

4.8 MSD TASKS—INFORMATION AND CONTROL

A manufacturing/supply operation can only be controlled effectively if the machines, operators and managers have the means to communicate to each other effectively. The MSD tasks in this area will deal with the analysis and specification of the organized data structure within an MS information system (MIS). The major tasks include design of the databases, the selection and location of hardware and software, and the selection of managerial roles which will be responsible for certain decision centers. The key requirements for the complete definition of an MS information system include the specification of process/functional structure, data structure, dynamic sequence of data, and cross-checking to ensure system integrity. Accordingly, the following tools can be used for the design tasks in this area:

- A *function diagram* (such as relatively high-level $IDEF_0$ models) to define the functions involved in various operational areas.
- A *data flow diagram* (DFD) to specify the data flows into and out of these functions, as well as data links within the functions.
- A *logical data model* (LDM) to identify the relationship between data entities.
- An *entity life history* (ELH) to specify the life sequence of an entity, if required.

Process/functional specification: data flow diagram

Function diagrams, such as those specified by the IDEF model of the MS system, are normally used first to specify the functions involved in the various operational areas of an organization. Having established such a functional hierarchy, it is then necessary to examine the data required for their operation, frequently by using a DFD to show:

- what data are needed to perform the functions,
- how data enter and leave the functions,
- where the data are stored,
- which functions generate changes of the data, and
- who provide, use and modify the data.

A0 Level Name:	A0 Level Definition			
Narrative : A0 MAKE AND DELIVER TO CUSTOMER ORDER				
A1 Level Functions	Description	A2 Level Functions	A3 Level Functions	Comment
A1	FORMULATE MANUFACTURING SERVICE PLAN	A11 Sale & Contract		
A3	GATHER RESOURCES		...... A322 Acquire material & bought- out items	

Figure 4.20 Example of an MS functional structure

For instance, suppose from the function hierarchy of the example order-handling MS system, the function blocks highlighted in Figure 4.20 are identified as the key functions to the processing of customer orders. Then a DFD of order-processing may be developed as shown in Figure 4.21. As can be seen, a DFD is a functional picture of the flows of data through the system. Similar to IDEF modeling, their development also follows a top-down process. Thus, a high level DFD can be developed into its lower levels of decomposition.

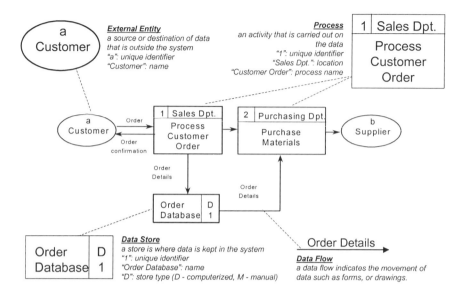

Figure 4.21 Example DFD model derived from an MS function

Data structure definition: logical data model

Next, an LDM is required to specify the data requirements of the system. The relational data model based on entity-relationship diagrams is perhaps the most widely adopted approach for this purpose. Such data representation uses the following concepts:

- *Entity*. An entity is considered to be anything about which the company wishes to store information. Examples of an entity include: an employee, a department, a supplier. An entity is shown as a rectangle containing the name of the entity, and normally an  identifier which provides a unique key to help identify a specific instance of that entity.

- *Relationship*. A relationship describes the mutual relationship between the entities, represented by a diamond containing a name. Participants of a relationship are connected to it by  straight lines, each labeled with *one* (a straight end or "1"), or *many* (a triangular end, an "*m*" or "∞") to specify whether the coexisted relationship is one-to-one, one-to-many, or many-to-many.

The cross-checking rule between MS functions/DFDs and the LDM as shown in Figure 4.22 can be used to identify the data entities of the system. For example, since the top-level function/DFD data sources or data stores normally indicate the existence of a data entity, the sample DFD diagram of order-handling reveals three data entities: *Customer*, *Order* and *Supplier*.

No	Entity Name	Attribute 1 (Key)	Attr 2	Attr 3	Attr 4
	Customer	ID	Name	Addrs	....
	Order	ID	Prod.	Qty	....
	Supplier	ID	Name	Addrs	....

Figure 4.22 Identification of entity-relationships

Following this, the relationships amongst them can be established through a simple entity matrix as shown in Figure 4.22 (*Worksheet MSD I.2*). With this matrix, each relationship can be established in turn. As can be seen, if two entities are related, this is entered in the cross box between them. If required, the nature of the relationship can be further specified, resulting in the LDM diagram shown in Figure 4.23. The relationship in this model reads: a customer may place one or more orders; an order is placed by a customer, one or more orders are placed with the supplier; a supplier supplies at least one order.

Figure 4.23 Sample entity-relationship model of the MS function

In addition, a set of attributes is used to specify the properties of a data entity. A relational data structure presents data entities in tables that specify their natures through these relevant attributes. Each table has a "key", which is a piece of data that uniquely defines a given data set. Relationships between these attributes are then used to link related entities. An example of a more complete MS logical data structure is shown in Figure 4.24, in which each data entity is identified by a key attribute (in bold), and the specific properties of each entity are defined by its own attribute set. In addition, the logical relationships amongst the entities are also clearly specified.

A logical data model, such as the example shown, here presents the conceptual structure of an MS database. When combined, the two sets of models (DFD, LDM) can provide a relatively complete presentation of the structure, contents and operation of an MS information system. The integrity of a system can be guaranteed by cross-checking between these parts using the rules summarized in Figure 4.3. An information system thus specified can be implemented in practice by using commercially available relational database management systems (DBMS).

Such software systems provide tools for relational data management such as data table definition and manipulation, as well as user-interface development.

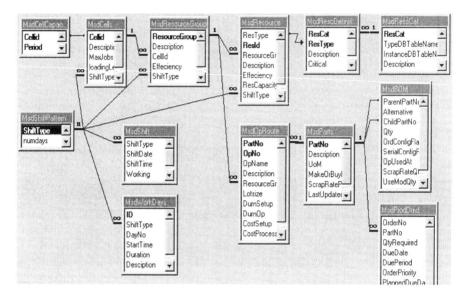

Figure 4.24 Example LDM model of an MS operation

Analysis and development process

Using the modeling techniques outlined above, the process normally followed for the analysis and development of an MIS is outlined as follows (Figure 4.25):

- **Feasibility study**. In close relation to the system function model, an initial level-1 (top level) DFD of the system is created, including a description of each function. An overview LDM is also created. Together, these assist project planning by identifying which areas need to be investigated. If desired, this DFD is decomposed to the next level, again according to the system function model. Following the problem-solving cycle, a number of outline options are formulated, from which one will be selected for further development.
- **Analysis**. The conceptual level DFD is then decomposed to lower levels as necessary. This leads to the specification of system LDM. Sources within the required system are analyzed to produce data-grouping, and to show their relationships. The DFD and LDM are used to validate one another by following the cross-checking rules.
- **Specification of requirements**. The system is next made more logical by showing what is to be achieved. All user requirements and functions should be considered for their relevance to the system being designed, and all the required data are included. Based upon the selected option, a new system specification can be created. The LDM will have to be updated to ensure that all the required data are available. The specification of requirements is expanded to give detail necessary

to build the system. Dialog design is used to chart the interactions between the system and the operator.

- **Selection of IT environment**. At this stage there will be enough knowledge for the designer to select the hardware and software environment for the system's development and implementation.
- **Physical design**. The logical data and processing designs are converted into a design which will run on the selected environment. Cross-checking should again take place to ensure the system's completeness, management and operational support.

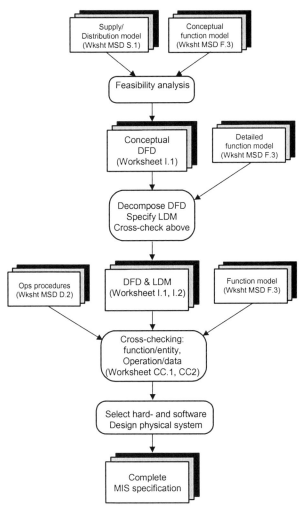

Figure 4.25 General processes for complete MIS specification

4.9 MSD TASKS—HUMAN AND ORGANIZATION

Without any doubt, this is one of the most important areas of manufacturing and supply systems management. Humans, and the way they are organized, are what operates and manages the actual transformation processes of the system, and eventually determines its success or failure. The keys are: *the right organizational structure, the right work/job system, and the right people for the job.*

Integrated human resource planning and management

The above core principles may look simple in printing, but they are perhaps the most difficult ones to achieve in practice—technologies and other hardware in a system are easily transferred, but human resources and organizational synergy are hard to copy. Also, in reality, the design and implementation of the organizational and human systems cannot be clearly separated. It is desirable to make the management of the new system responsible for its implementation right from the beginning. That is, the user should be made the system/process owner. This ensures user commitment to changes and to the success of the implementation in the long term. In this regard, the MSD team's role is to help the process by explaining the strategic requirements, the objectives, and the system designs to management and their teams, and by undertaking project management and coordination of implementation activities.

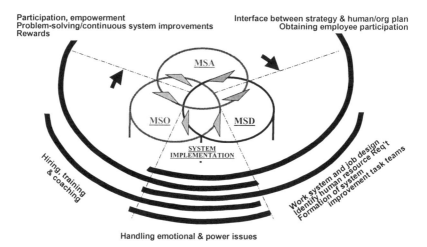

Figure 4.26 Integration of human resource processes within MSM

Integrated management of human resources is required within the MSM context, with the aims to provide, coordinate, motivate, and empower people at all levels and in all functions, so as to effectively support the organization's strategic needs. Identification of the right organizational structure and work systems through MSD tasks is therefore only a part of the whole picture of human resources and change management that must be logically interwoven within the overall MSM structure, as shown in Figure 4.26. The following are the main activities involved in this domain.

Interface between MS strategy and human resource needs

The human resource plan of an organization should be aligned with its strategic requirement. The key considerations of alignment identified here are concerned with the links between MS strategy and organization structure, employees, training and learning, and appraisal and reward.

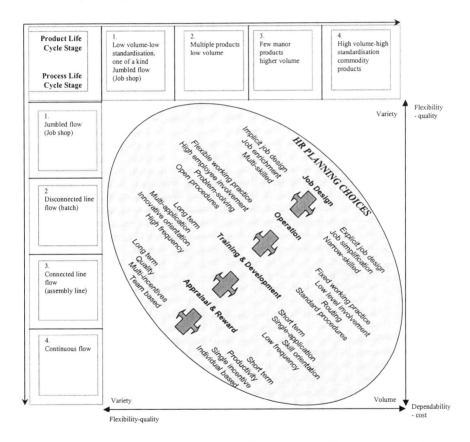

Figure 4.27 Relationship between MS strategic requirement and human resource need

It has been suggested that most organizations will find their strategy belonging to one of a few strategic groups, as listed in *Tool/Technique MSA 4.3.1*. When combined with the product/process matrix, this provides some general guidelines for the alignment of human resource plans with company strategy. The more traditional practice of human resource management, as reflected by the choices at the bottom-right section of Figure 4.27, supports a cost-reduction strategy ("caretakers"). This normally involves mature products with relatively long product life cycle in a stable market. On the other hand, an innovation and/or quality strategy needs the support of human resource plans that are grouped towards the top-left section ("innovators"). These help to develop a set generic profiles to provide guidance to help cross-check human resource and strategic requirements. Depending on the particular type and stage of the MS operation (and its current positioning), certain human resource plans may be more appropriate. Therefore, if the state of the enterprise is known, then appropriate human resource needs may be suggested for consideration according to the compatibility of manufacturing strategies with respect to the organizational state and resource deployment, as outlined in Figure 4.27. In such a way, both the strategy and its supporting structure might progress and develop in a consistent and logical manner.

Regardless of the type of operation, in order to develop effective organizational structure and human resources, it is important for an company to gain the participation of the entire workforce. It must fully appreciate the value of the skills and the experience of its employees. In addition, participation is crucial to avoid resistance to change and to ensure that the changes brought about by an MSD project will last. Some conditions are necessary to obtain participation from the workforce, including:

- *Communicating.* Strategy and objectives should be communicated to the workforce at all levels of the organizational hierarchy. A high level of awareness of the aims, goals and changes should be maintained by everyone involved.
- *Facilitating, not authoritarian.* An environment should be created in which freedom and flexibility enable the staff to make the best use of their creativity, expertise and skills. Also, taking risks and making mistakes should be allowed. This will increase the output of the staff in terms of idea creation and innovation.
- *Following suggestions.* It is important to make sure that all the ideas and initiatives generated by the workforce be taken into consideration. Every suggestion should get a response, and a bonus system could be instigated to reward the best suggestions.

Work system and job design

Work system design is concerned with how employees are organized into both formal departments/units, and informal work teams. Job design refers the definition of individual responsibilities. It is essential that roles, behaviors and responsibilities of all positions in the organization be defined and/or reshaped prior to the implementation. Two main activities are therefore (see Figure 4.28):

- Analysis of the business and human resource plans, which indicate the types of skills and competencies that may be required in the future, and the number of people with those skills that will be needed.
- Job analysis to examine in detail the content of the jobs and what knowledge and skills are required of the jobholder.

It is also necessary to decide how to acquire the new skills needed. This may involve additional training of existing employees or recruitment of skilled people outside the company. Either way, changes inevitably occur whenever an MSM cycle is initiated and followed through. People react differently when facing change, and change can be difficult if the emotional dimension of the employees is not managed. In order to increase the proportion of staff with a positive reaction, it is necessary to work on the following principles:

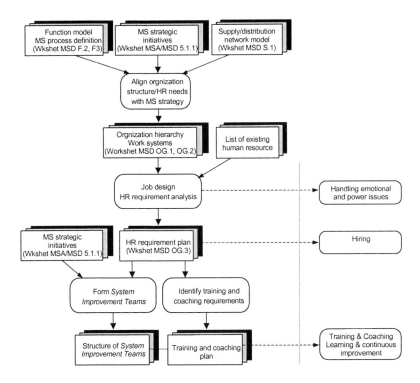

Figure 4.28 General processes within the *human & organization* MSD domain

- *Analyzing the current situation.* The organization must evaluate how the business and its employees are likely to react to change. Information and criteria for this may be directly extracted from the results of the relevant MSA worksheets. Also, assessment can be carried out by using personal interviews, with each key person and with representatives of the whole workforce.
- *Identifying and dealing with resistance.* The nature of change makes demands on the employees in term of augmenting their technical and social skills, their ways of thinking and their attitudes. The earliest possible involvement of the employees concerned will help prevent or diminish resistance. Also, management should provide continual reaffirmation of a will to successfully carry out changes, and should endorse the workforce's efforts and the results they obtain.
- *Handling power issues.* In any change process, it is vital to gain the support of those who hold power in the organization. Handling power issues begins with an analysis of the level of support for change. The power exerted by key staff and people with unusual skills must be recognized. Once the level of importance of

each person in power has been assessed, the mapping of the power structure can be made against the changes planned. This analysis identifies areas of strength and areas at risk. Recruitment and training may be necessary to fill in the power/skill vacuum. System structures and procedures should be specified accordingly in order to establish the new power situation quickly and effectively.

System improvement and learning

It is now a universally accepted, and frequently enforced view that continuous improvement and learning should be treated as an integral part of the organizational culture. Fundamentally, this view has its roots in systems thinking. It reflects the feedback requirement of any open system that needs to adapt to environmental change or to achieve new goals (see Section 4.4.3). In fact, it should be clear that the structure and approach of this MSM framework is designed precisely for the purpose of helping an MS organization become agile and adaptive, through the learning mechanism that is embedded in the MSA-MPM cycle, and through the execution of MSD projects that are aimed at continuous improvement.

Therefore, one of the key requirements for a successful MS operation is to institutionalize continuous improvement and learning, so that they become an embedded part of the daily work activities of all employees. The establishment and empowerment of *system improvement task teams (SITs)* provide a mechanism to put this into practice. Their structure and aim should resemble that of a racing team, with multi-skilled team members to fine-tune and continuously improve the system's structure and performance, so as to keep the "MSM driving wheel" rotating towards the desired strategic direction. An SIT team should consist of a team coordinator, and a number of mixed employees belonging to different departments or units in the organization, at different levels of the organizational hierarchy. They present a cross-functional task force that meets to carry out problem-solving tasks related to various issues of system improvement. The life-span of a SIT team depends on the tasks in hand. Some teams may be formed to deal with a specific problem and are disbanded once the task has been completed. Others may be more enduring, dealing with ongoing issues of both an operational and system-related nature. The character of an SIT team may be determined according to:

- *Management task teams*: consisting of managers from various departments. Its role resembles that of a committee with the responsibility to plan, coordinate and track the progress of the current SIT teams.
- *MSD task teams*: a project team formed specifically to develop a new system function, by accomplishing the MSD tasks as previously specified. Led by a principle function owner, the team should consist of the designers as well as members of the owner function, the customer and the inputting functions. It is good practice for this team to be responsible for the design, implementation and operation of the system function(s) in its area of responsibility.
- *Quality circle teams*: consisting of a small group of managers and workers from one or more functional areas. Such a team meets periodically to identify, analyze and solve problems in order to improve quality and productivity.

Training and coaching

Change demands an upgrading of employees' knowledge and skills. The nature of their activities may transform drastically, making it necessary for workers to acquire new expertise. On the other hand, these same employees will also be in the front line of the change process, needing to know how to carry out change. Change also requires alterations in the way people behave, and training alone cannot achieve that. It may prove necessary to give some people in the organization (especially managers) one-to-one support to help them accept change and transform their methods and behavior in line with the objectives defined in the vision. This type of support is known as coaching. All these approaches help human resource planning and management to align with the MS strategy. When designing training and coaching programs, therefore, the following points need to be considered:

- *Objectives of the training program.* Training objectives should be linked to the MS strategic initiatives.
- *Content and frequency of training.* Training plans should be based upon job design and skill requirements, determined by what the trainee should be able to do after completion of the training.
- *Who and where.* Can training be provided by managers, team leaders, colleagues in the company or only from others outside the company?

Creative use of information technologies such as CBT can both speed up the training process and increase its quality. With such approaches, online and on-demand training can be made possible, providing comprehensive "know-how" on the processes involved, and addressing the need for the timely provision of training about specific tasks. These approaches have the potential to help a first time operator/manager learn how to carry out a new task/operation from start to finish with either minimal or no external training. Typically, it provides *structured lessons* that guide a trainee through a training sequence, *operational simulation* with a virtual environment for the trainee to explore and experiment through simulation, and *learning assessment* to check the trainee's progress. A case of its application is presented in Chapter 7.

4.10 CROSS-CHECKING

The results produced from the six MS design areas provide a relatively complete presentation of the structure, contents and operation of the MS operation under consideration. However, the integrity of the system design needs to be assured. This can be achieved by cross-checking the designs at both the conceptual level and the detailed levels using *Worksheet MSD CC.1* and *CC.2*, respectively. The general cross-checking rules have been presented in detail in Section 4.3. An example of system level cross-checking is given in Figure 4.29.

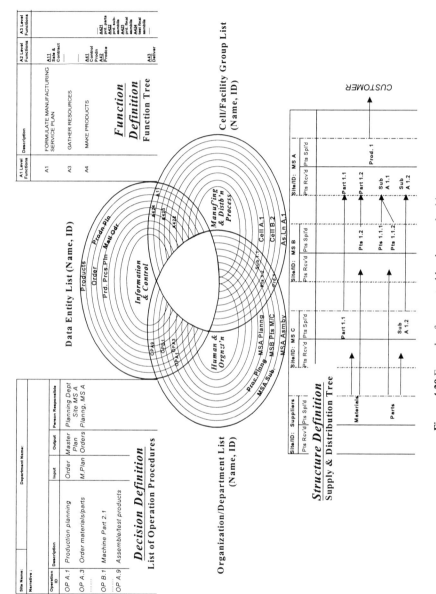

Figure 4.29 Example of conceptual level cross-checking

Task Document MSD 1—Execution of MSD Tasks

TASK OVERVIEW

Having identified the relevant MSD tasks and specified their evaluation criteria and constraints, the actual execution of a task involves a problem-solving cycle, typically consisting of the following:

Synthesis of design options. This generates a number of possible alternative design concepts. The features and characteristics of these design concepts/options should be clearly documented.

Static analysis and evaluation of options. Static analysis is the first-cut step aiming to establish whether the options will, in general, satisfy the requirements. Analysis is based on static criteria such as the overall level of capacity, or the overall number of parts required over a specific period. It consists of the following elements:

- Data collection. Identification and collection of the data required to carry out model-building and design concept evaluation.
- Model building. A model is normally used to predict the possible results of the individual design concepts. The types of models used at this stage are normally deterministic and steady-state, and frequently numerical (Tool/Tech. MSD 1.1). For example, a spreadsheet-based model is often used for high-level financial evaluation at the static stage.
- Evaluation of design options. Using the model developed, outcomes should be predicted for each of the design options. A comparison of these can then be carried out against the MSD task objectives. If some of them satisfy the objectives, one or more can be chosen as the candidate for further development.

Dynamic analysis and evaluation of options. This aims to refine the preliminary choices identified above by evaluating their levels of satisfaction under dynamic operating conditions. This step involves the same steps and procedures as above. However, the models used for this purpose are normally dynamic/non-deterministic in nature (such as computer simulation), and the criteria are often related to changes in performance: e.g., seasonal variation in demand or the dynamic level of work load in a machine shop.

Final evaluation and documentation. Evaluation of design options should be based on utility value analysis, and risk and sensitivity analysis. The final design, which presents the best option from the list, satisfies the objectives under both static and dynamic conditions, and with the minimum risk.

(left margin label: TASK DESCRIPTION)

TASK LINKS   *POSITION IN MSM FRAMEWORK*

INPUT FROM :

MSA/MSD 2.3 (MSD list and objectives)

Other relevant MSD tasks within the identified project.

OUTPUT TO :

Immediate results: other relevant MSD tasks within project.

Final results: MS system implementation.

OUTPUTS

System design from the required project(s), evaluated, detailed, and ready for system implementation.

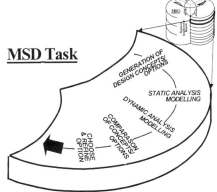

MSD Task

GENERATION OF DESIGN CONCEPTS/ OPTIONS

STATIC ANALYSIS MODELLING

DYNAMIC ANALYSIS MODELLING

COMPARISON OF CONCEPTS/ OPTIONS

CHOOSE & REFINE OPTION

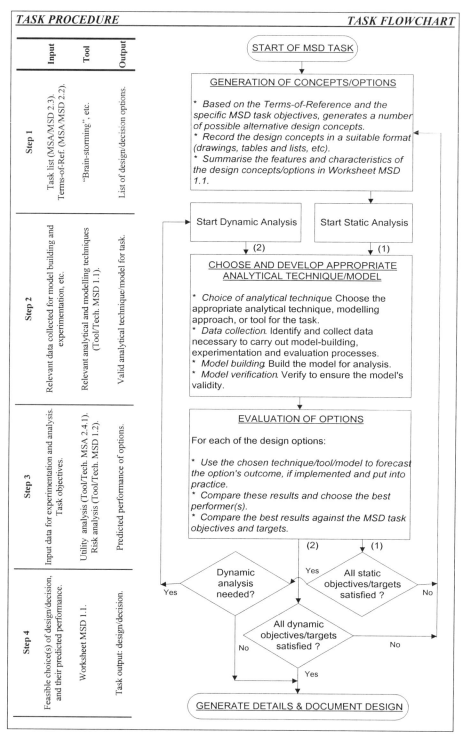

TOOL/TECHNIQUE MSD 1.1—MSD Analytical Tools

Model and problem type

The types of modeling techniques used in evaluating alternatives are highly diversified, including mathematical, physical and simulation. Also they can be either quantitative or descriptive, as summarized below.

MODEL TYPE	Descriptive	Physical	Mathematical	Simulation
Prediction Method	judgment	manipulation	mathematical	simulation
Optimization Method	judgment	experiment	mathematical analysis	experiment
Cost	low	high	medium	high
Ease of Communicatn	poor	good	poor	excellent
Limitation	not repeatable	cannot represent information process	can only cope with simplified cases	optimal solution not guaranteed

The choice of the modeling approach should consider the characteristics of the system in question and the nature of the problems to be tackled. It is also dependent on the type of problem, nature of performance measure and the objectives. Other factors, such as the amount and type of quantitative information available, the amount of time and money at the analyst's disposal and the facilities available (such as computer hardware and software) should also be taken into consideration. In general, analysis in MSD can be classified according to the following groups:

- *Static vs. dynamic*. Static analysis refers to a situation where the problems and system parameters involved are fixed in nature. A typical example is the evaluation of the average capacity level required over a relatively long period of time, based on the estimate of the overall demand in that period. When the problem is dynamic, on the other hand, the time dimension must be taken into consideration.
- *Deterministic vs. stochastic*. Models of this type are based on an algebraic relationship. This kind of model is used when the mechanisms governing behavior are understood and thus, the relationships among system variables and parameters become transparent to the model builder. In a stochastic situation, the governing mechanism is not totally understood. Therefore, the relationships among the system parameters—or sometimes the values of system parameters—can only be estimated through techniques provided by statistical analysis.

Mathematical techniques

A number of typical mathematical techniques are outlined below. Details concerning theory and application can be found in many technical papers and textbooks.

- *Linear programming (LP)* is a mathematical analysis technique applicable when it is necessary to find the optimal value of a linear effectiveness function subject to a number of linear constraints. It has found many applications in the MSD area, since many problems lend themselves naturally to LP models. These include, for example, the problems of capacity analysis, site allocate, facility layout and line balancing. Also, the models constructed are relatively easy to understand. Computer software packages are readily available so that solutions of the models can be easily obtained. What-if analyses can also be carried out.
- *Time series (TS) analysis* typically utilizes historical data to statistically predict a future level, such as the forecast of demand for a product. This approach involves a number of mathematical models such as *linear regression* analysis, the *weighted moving average method*, and the *exponential smoothing method*. The choice of the model depends on the

nature of the market and the products concerned (trend, seasonal and cyclical variation, etc.).

- ***Queuing theory (QT)*** is useful because queuing is a predominate feature of any MS system where parts or customers wait for the service from certain resources. The operation of a machining station is a good example. When a work piece arrives for machining, it will wait at a buffer store if the required machine is not free. Once the machine is ready to take a job and there is a job waiting in the buffer, the machining process can start. Queuing theory is the study of such a situation and may provide calculation of the average values about such system measures as the average waiting and throughput times of parts, and the average utilization of resources in the system.

- ***Inventory models (IM)*** are used to find the best inventory structure and policy regarding system parameters such as inventory level, lot size and reorder point. The most common model in this category is economic order quantity (EOQ), which calculates, under simplification of the situation, the optimal level of the inventory level that minimizes the total inventory costs.

- ***Cellular formation (CF) algorithms*** have been developed specifically for the MSD task of grouping parts and machines into mutually independent cells. Such a layout has the advantage of simplifying production control and material flow in the system. A cellular-based MS operation is the basis for many modern approaches such as group technology, JIT, total quality management (TQM), learn/agile manufacturing and FMS. Cellular formation algorithms can be classified into two main types: the product-based and the product/process-based methods. The product-based methods arrange parts into groups according to the shape and size, while the product/process-based approach also uses production information to carry out the analysis.

- ***Analysis hierarchy process (AHP)*** is a general technique for dealing with multi-variable decision-making. It involves both tangible and intangible system parameters. With this approach, an objective is initially set for the required solution. This is then split into sub-objectives, which must be fulfilled to achieve the higher objective. Intangible benefits can be included in the formula through the use of a utility vector that allows the evaluation of alternatives.

- ***Other approaches*** include ***Petri nets (PN)*** which were originally used for the analysis of computer systems, but are being used increasingly in MSD areas. Like queuing theory, this technique is suitable when interdependent components must interact to initiate a certain event. It can therefore be used to analyze the material or information flows within an MS operation. ***Neural-nets (NN)*** are based on our current understanding of the structure and working of the biological nervous systems. Specially built neural-nets models can utilize fuzzy information and can deal with problems such as design optimization, group technology and cellular formation. In addition, numerous ***heuristics methods (HM)*** have been developed for various types MSD analysis. Although not always strictly mathematical, they normally are theoretically based.

Computer simulation

Computer simulation is fundamentally an experimental approach for studying certain functional properties of an organization by experimenting with an appropriate computer model rather than with the actual system. As far as MS system design is concerned, computer simulation frequently provides a flexible and powerful technique compared with the others. It is one of the most effective tools available, particularly as a method for evaluating the dynamic characteristics of a proposed solution. With a properly constructed computer simulation model, a system designer may experiment with different manufacturing runs, new operational conditions, new layout of equipment, different cycle times, etc. This allows the designer to predict how the system will perform when put into operation. The most relevant type of simulation in the MSD area is known as discrete-event simulation. In contrast to continuous simulation models, which are usually based on certain mathematical equations (as exemplified by the so-called *System Dynamics* approach), discrete-event simulation is concerned with the modeling of a system by a representation in which the state variables change at sudden distinct events. For instance, the state of a machine changes discretely from one state to another at a certain point in time. The time taken for the machine to process a work piece can either be sampled from some appropriate distribution (random simulation), or set to a known constant (deterministic simulation). This is also true for other activities in the system. Therefore, referring to these known activity times can simulate the next change in the state of the system. During this operation, a work piece would enter the system and wait in a queue for its turn to be processed by the machine. When the work piece is at the head of the queue and the machine is ready to take on another job, it will be taken from the queue and loaded on the machine so that the machining operation can take place. When the machining operation is completed, the value of certain attributes of the work piece will have been altered due to the transformation associated with this activity (in this case, a raw material machined into a part).

According to the above, simulation software generalizes the necessary simulation procedures and makes the programming of simulation model a relatively easy task, by providing:

- graphic interface for model construction,
- event and simulation time handling,
- graphic animation of the processes involved,
- interactive control of simulation processes, and
- results analysis and report generation facilities.

Once a model is built and validated, experiments can be carried out to simulate the system behavior under different operating conditions or with alternative system configurations.

Estimation of means and variance of the response of a model under a particular set of inputs and operating conditions is of particular importance to simulation study. This is because the mean value of individual observations is often used as the system performance criterion in many MS problems. Such problems may include the mean system throughput time, mean machine utilization, mean order tardiness and mean work-in-progress level. Comparison of these predicted outcomes enables the analyst to choose from alternative solutions. As an aid to decision-making, the technique of computer simulation has many desirable features including:

- *Flexibility.* Once a model is developed, it can be modified to include new features to evaluate additional alternatives.
- *Study of transient behavior.* When analyzing the dynamic characteristics of a system, computer simulation is frequently the optimum analytical approach.
- *Communication.* The ability to animate system behavior allows for ease of communication amongst designers, and between the designers of the system and its user. This makes the user actively involved throughout the system design cycle.

However, it should be noted that the amount of time and expertise required to construct a simulation model can be significant. Also, decision-making using simulation is by nature through statistical experimentation. Optimal solutions are not always guaranteed.

The table on the next page provides an overview of a number of approaches used to model MSD problems, together with some of their typical applications in specific MSD areas.

Examples of analytical techniques and their MSD applications		
MS Design Area	**Model application**	
	Mathematical	**Computer Simulation**
REQUIREMENTS — System Function	LP: aggregate capacity planning TS: demand forecast IM: make vs. buy AHP: process selection	Market study and demand forecast, total capacity planning, financial evaluation in this and the following areas.
System Structure	LP: site location CF: parts/sites grouping QT: order throughput time, overall site capacity NN: parts/sites grouping	SD: supply-distribution structure
System Decision	PN: decision network and structure	Simulation of decision processes and functions.
CONCEPTUAL DESIGN — Manufacturing and Supply Process	LP: detailed capacity planning, facility layout, line balancing, factory/warehouse location QT: as above plus throughput time analysis, capacity utilization CF: cellular formation NN: cellular formation AHP: process selection	System-wide conceptual design and specification: inventory and capacity planning, master production scheduling, evaluation of other production management decisions, and specification of MS facilities and layout at the conceptual level.
Human and Organization	LP: resource planning QT: resource planning CF: cellular formation	As above, but dealing with human resource requirement planning.
Information and Control	Petra Nets: computer networking	
DETAILED DESIGN — Processes	LP: line balancing, material-handling, facility layout QT: as above plus throughput time evaluation and resource utilization IM: make vs. buy AHP: machine selection	Detailed planning and specification of MS systems in this and all the following areas such as: planning of detailed work loading, study of scheduling and job issuing policies, identifying bottlenecks, facility layout and material handling.
Facilities	LP: facility layout	
Supports	QT: support facility capacity TS: maintenance policy	
Planning	LP: production planning QT: evaluation of planning policies	
Control	LP: scheduling IM: lot sizes, batch sizes PN: robotic path layout (collision detection) AHP: hardware/software selection	
Human	LP: capacity planning CF: operator assignment	
Warehouse and Transport	LP: facility and warehouse location IM: lot sizes, batch sizes, inventory level TS: inventory level QT: throughput & queuing time	

TOOL/TECHNIQUE MSD 1.2—Evaluation of Design Alternatives

Each design option should be evaluated using techniques of utility value analysis and risk management. For details about structured decision-making using the utility value analysis, please refer to *Tool/Techniques MSA 2.4.1.*

Risk management

Risk management is concerned with what investment and work might be at risk if the project is delayed and/or abandoned, and what course of action

should be undertaken if such events occur. It involves three major components: (1) identification of risks; (2) prediction of effects of risks; and (3) creation of contingency plans. Risks arise mainly because of uncertainties involved in the financial, business, social and natural environments. A number of techniques can be used to locate these risks including, for instance, the *herringbone diagram* which relate the effect of a failure to the elements that lead to that effect. For example the risk of going over budget may be attributed to three major categories: labor, rework and suppliers, as shown above.

Once risks have been identified, they can be analyzed by giving each risk two values on an appropriate scale: a likelihood value, to measure how something will go wrong, and an impact value, to reflect its effect on the project. The multiplication of the two values provides a weighting for each risk factors. Additionally, more sophisticated mathematical analyses, such as probabilities and simulation techniques, may be used for the same purpose. For example, the single estimation of time requirement for each activity along the critical path may be replaced by a proper probability distribution. Through simulation, the overall project duration can be estimated, together with its probability value.

The analysis can be further enhanced by carrying out sensitivity analysis, with the aim to see how design solutions vary as a result of changing parameter and constraint values. For example, constraints may be relaxed, measurement standards may be varied, objectives may be altered, and even project scope may be expended or contracted. By holding all other factors constant, the sensitivity of assumptions can be evaluated in terms of total cost and system effectiveness. The information will allow the creation of contingency plans that highlight the actions needed to minimize the possible undesirable effects.

Investment appraisal

This is concerned with the financial appraisal of manufacturing/supply technology. Advanced MS technologies and equipment are becoming more complicated and expensive. This has made acquisition of the necessary investment capital difficult, due to the fact that the traditional methods of investment appraisal (e.g., payback or discounting cash flows) demand rapid repayment, while the new technologies are increasingly infra-structural, and provide long term rather than short term benefits. Examples of these intangible gains, which can be as important as the tangible ones, include: ability to respond to the customer consistently and predictably, rapid response to market change with respect to product volume, product mix and product change, shorter product lead-time, reduced inventory, improved manufacturing controls, real-time control of components, better quality, high utilization of key equipment, reduced tooling, simplified fixture design, reduced direct labor content, reduced fitting and assembly requirements, reduced overhead cost, new disciplines being added to the planning process, and the possibility of integrated manufacturing/supply. These improvements can be transformed to improvements in profits through, for example, better figures of sale, reduced inventory costs and reduced operating costs. The AHP technique can be used to provide an alternative to the traditional financial analysis.

WORKSHEET MSD 1—MSD Task Execution

Project Title:

Person(s) Responsible:

Version: **Date Completed:**

MSD project: *MSD task*:
MSD type: static/dynamic

SUMMARY OF DESIGN CONCEPTS

Alternative	1	2	3
Description			

GENERAL CHARACTERISTICS OF DESIGN CONCEPTS

Key Features			
Level of Complexity			
Level of Skill Required			
Costs			

DETAILS OF DESIGN

Drawing No.			
Data Sheet No.			
Other Documents			

DETAILS OF ANALYSIS

Techniques And Models				
Details of Experiments (Aims and Approaches)	1			
	2			
	3			
	4			
	5			

RESULTS OF ANALYSIS

Results of analysis/ Experimentation	1			
	2			
	3			
	4			
	5			
Risk Analysis				
Sensitivity Analysis				
Weighed Overall				

WORKSHEET MSD F.1—*Product List*

Project Title:

	PRODUCT STRUCTURE/PARTS LIST			

	SYSTEM	DESCRIPTION	DWG NO	REV
AUTHOR:	DATE	SHEET	OF	

ITEM		PART NO. QTY./YEAR	
DESCRIPTION			

Assembly No.	Description	Sub - Assembly No.	Part No.	Comment

WORKSHEET MSD F.2—*Production Process Definition*

Project Title:

PRODUCTION PROCESS DEFINITION

	Produc/ Parts	DESCRIPTION	Name	ID
AUTHOR:				
	DATE	SHEET	OF	

Op. No.	Activity ID	Activity Nme	Cell/Machine ID	Cell/Machine Nme	N/C Program / Process Data Narrative	N/C Program / Process Data ID	Unit Time	Unit Cost

WORKSHEET MSD F.3—Definition of Functional Structure

Project Title:

	SYSTEM FUNCTION STRUCTURE			
AUTHOR:	SYSTEM	DESCRIPTION	DWG NO	REV
	DATE	SHEET	OF	

A0 Level Name:	A0 Level Definition

Narrative :

A1 Level Functions	Description	A2 Level Functions	A3 Level Functions	Comment

WORKSHEET MSD S.1—Supply and Distribution Modeling

Project Title:

SYSTEM SUPPLY & DISTRIBUTION STRUCTURE				
AUTHOR:	SYSTEM	DESCRIPTION	DWG NO	REV
	DATE	SHEET	OF	

Site Name/ID:		Site Name/ID:		Site Name/ID:		Site Name/ID:	
Pts Rcv'd	Pts Spl'd	Pts Rcv'd	Pts Spl'd	Pts Rcv'd	Pts Spl'd	Pts Rcv'd	Pts Spl'd

WORKSHEET MSD D.1—List of Operations Procedures

Project Title:

	LIST OF OPERATIONS PROCEDURES			
AUTHOR:	SYSTEM	DESCRIPTION	DWG NO	REV
	DATE	SHEET	OF	

Site Name: **Department Name:**

Narrative :

Operation ID	Description	Input	Output	Responsibility

WORKSHEET MSD D.2—*Operations Procedures*

Project Title:

OPERATIONS PROCEDURE

	OWNER	DESCRIPTION		DWG NO	REV
AUTHOR:					
	DATE	SHEET	OF		

Op. No.	ACTIVITY		LINKS		CONTROL		Invty&Lrng Business	Customer Financial
	ID	Effct	Input	Outpt	Description	Time		

PRFMNC MEASUR

WORKSHEET MSD P.1—*Capacity Requirement Analysis*

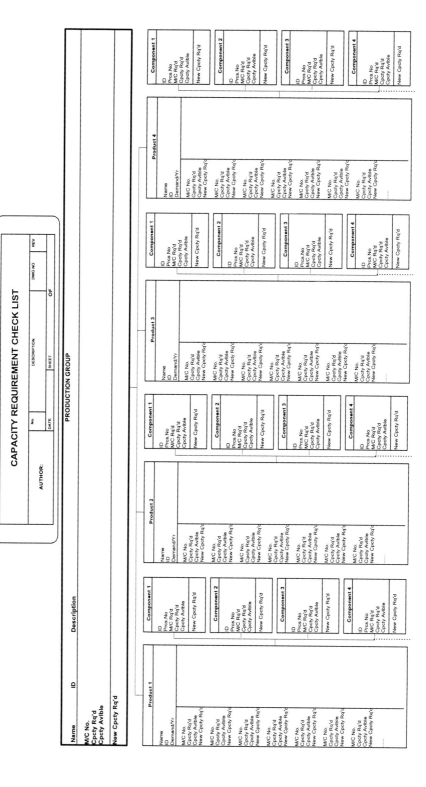

WORKSHEET MSD P.2—Cell/Facility List

Project Title:

MS FACILITY LIST					

AUTHOR:	SYSTEM	DESCRIPTION	DWG NO	REV
	DATE	SHEET	OF	

CELL ID.		CELL NAME	

DESCRIPTION

Equipmnt List (ID)	Description	Capacity	Cost	Comment

WORKSHEET MSD I.1—Data Flow Hierarchy

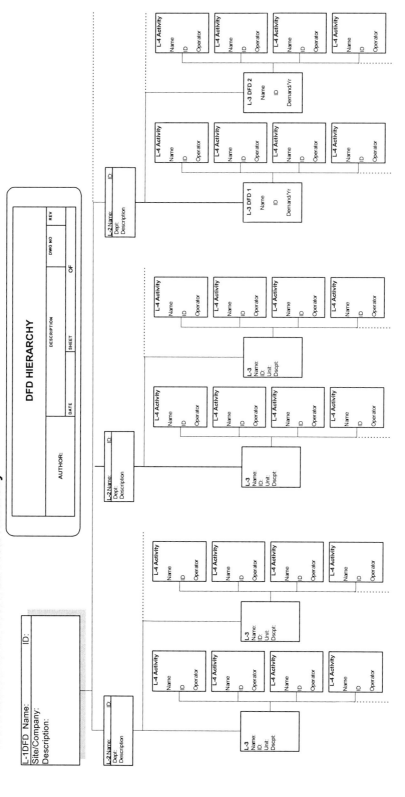

WORKSHEET MSD I.2—Structure of Logical Data Model

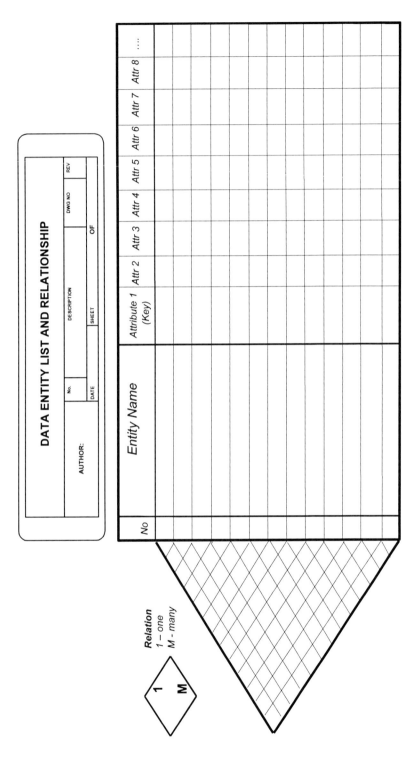

DATA ENTITY LIST AND RELATIONSHIP

AUTHOR:

No. | DESCRIPTION | DWG NO | REV

DATE | SHEET | OF

No	Entity Name	Attribute 1 (Key)	Attr 2	Attr 3	Attr 4	Attr 5	Attr 6	Attr 7	Attr 8	….

Relation
1 – one
M - many

1
M

WORKSHEET MSD I.3—Definition of Bill-Of-Material Structure

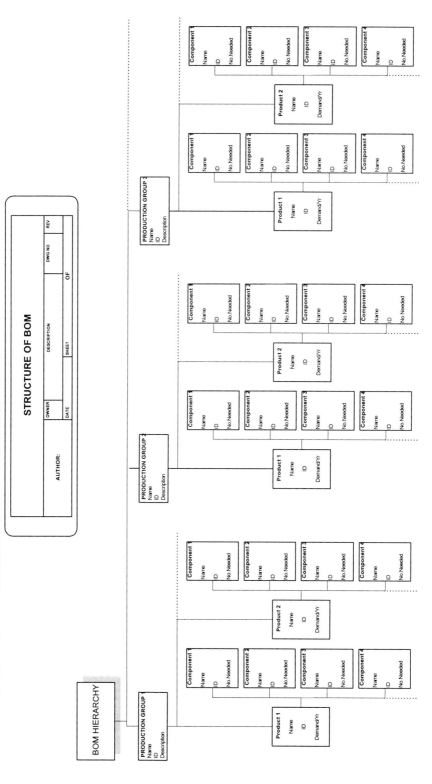

WORKSHEET MSD OG.1—Organization Hierarchy

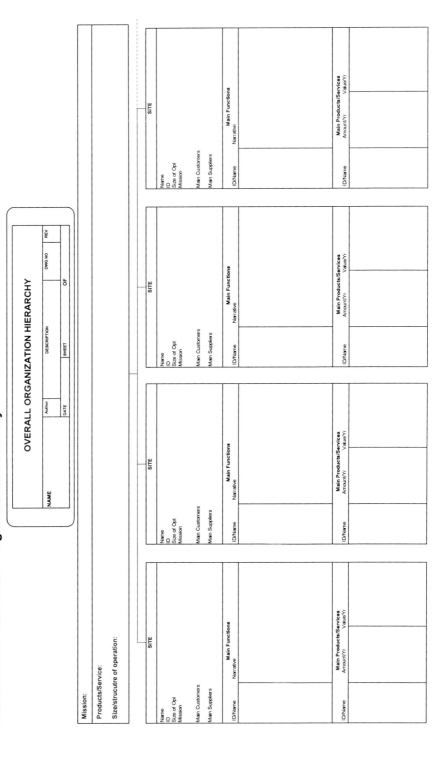

Mission:

Products/Service:

Size/strucutre of operation:

OVERALL ORGANIZATION HIERARCHY

NAME			
	Author	DESCRIPTION	DWG NO	REV
	DATE	SHEET	OF	

SITE

Name
ID
Size of Opt
Mission

Main Customers

Main Suppliers

Main Functions
Narrative

ID/Name

Main Products/Services
Amount/Yr   Value/Yr

ID/Name

SITE

Name
ID
Size of Opt
Mission

Main Customers

Main Suppliers

Main Functions
Narrative

ID/Name

Main Products/Services
Amount/Yr   Value/Yr

ID/Name

SITE

Name
ID
Size of Opt
Mission

Main Customers

Main Suppliers

Main Functions
Narrative

ID/Name

Main Products/Services
Amount/Yr   Value/Yr

ID/Name

WORKSHEET MSD OG.2—Hierarchy of an MS Site

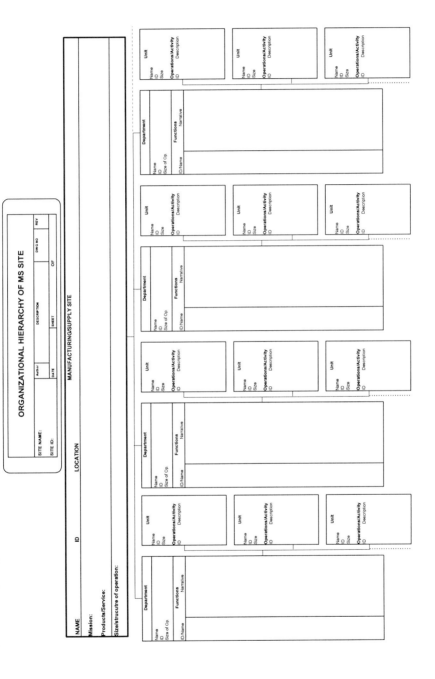

ORGANIZATIONAL HIERARCHY OF MS SITE

SITE NAME:

SITE ID:

Author	DESCRIPTION	DWG NO	REV
DATE		SHEET	OF

MANUFACTURING/SUPPLY SITE

NAME	ID	LOCATION

Mission:

Products/Service:

Size/structutre of operation:

Department
Name
ID
Size of Op

Functions
ID/Name   Narrative

Unit
Name
ID
Size

Operations/Activity
ID   Description

WORKSHEET MSD OG.3—Job and Human Resource Analysis

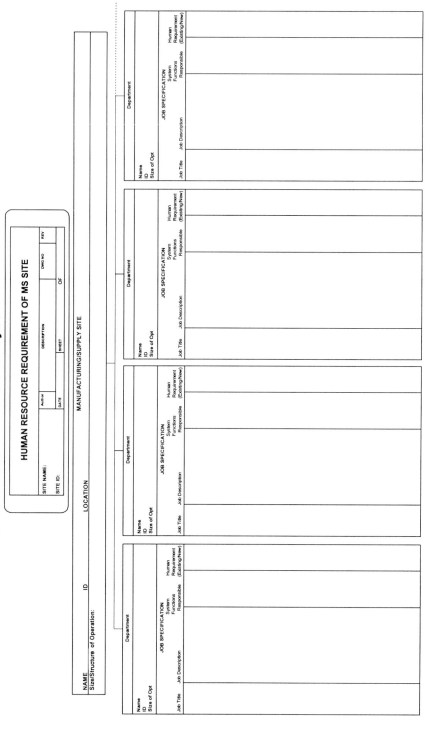

WORKSHEET MSD OG.4—Definition of System Task Teams

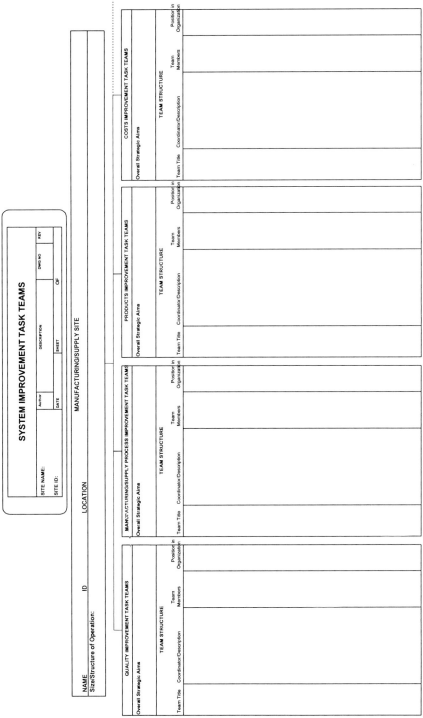

SYSTEM IMPROVEMENT TASK TEAMS

SITE NAME:

SITE ID:

Author | DESCRIPTION | DWG NO | REV

DATE | SHEET | OF

MANUFACTURING/SUPPLY SITE

NAME | ID
LOCATION

Size/Structure of Operation:

QUALITY IMPROVEMENT TASK TEAMS

Overall Strategic Aims

TEAM STRUCTURE

Team Title	Coordinator/Description	Team Members	Position in Organization

MANUFACTURING/SUPPLY PROCESS IMPROVEMENT TASK TEAMS

Overall Strategic Aims

TEAM STRUCTURE

Team Title	Coordinator/Description	Team Members	Position in Organization

PRODUCTS IMPROVEMENT TASK TEAMS

Overall Strategic Aims

TEAM STRUCTURE

Team Title	Coordinator/Description	Team Members	Position in Organization

COSTS IMPROVEMENT TASK TEAMS

Overall Strategic Aims

TEAM STRUCTURE

Team Title	Coordinator/Description	Team Members	Position in Organization

WORKSHEET MSD CC.1—Conceptual Level Cross-checking

Decision Definition
List of Operation Procedures

Site Name: _____ Department Name: _____

Narrative : _____

Operation ID	Description	Input	Output	Person Responsible

Function Definition
Function Tree

A0 Level Name: _____

Narrative : _____

A1 Level Functions	Discription	A2 Level Functions	A3 Level Functions	Comment

Data Entity List (Name, ID)

Cell/Facility Group List (Name, ID)

Organization/Department List (Name, ID)

- Information & Control
- Manuf'ing & Distb'n Process
- Human & Orgnzt'n

Structure Definition
Supply & Distribution Tree

Site Name/ID:		Site Name/ID:		Site Name/ID:		Site Name/ID:	
Pts Rcv'd	Pts Spl'd	Pts Rcv'd	Pts Spl'd	Pts Rcv'd	Pts Spl'd		Pts Spl'd

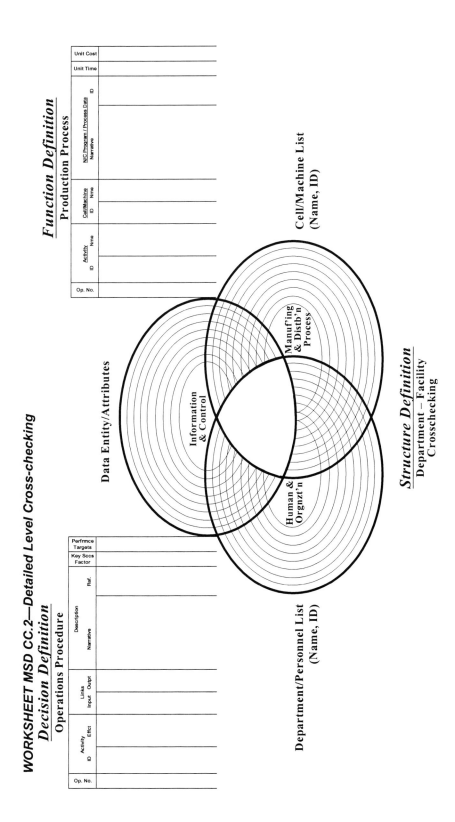

WORKSHEET MSD CC.2—Detailed Level Cross-checking

Function Definition
Production Process

Op. No.	Activity		Cell/Machine		N/C Program / Process Data	Unit Time	Unit Cost
	ID	Nme	ID	Nme	ID Narrative		

Cell/Machine List
(Name, ID)

Structure Definition
Department – Facility
Crosschecking

Department/Personnel List
(Name, ID)

Data Entity/Attributes

Information
& Control

Manuf'ing
& Distb'n
Process

Human &
Orgnzt'n

Decision Definition
Operations Procedure

Op. No.	Activity		Links		Description		Key Sccs Factor	Perfrmce Targets
	ID	Effct	Input	Outpt	Narrative	Ref.		

CHAPTER FIVE

MS System Implementation

5.1 INTRODUCTION

The production of a detailed system design is not the end of the story. Implementation of the design must consider the means by which the systems can be put into practice, while causing as minimal a disruption as possible. This stage involves planning for the implementation, seeking approval and physically making the necessary installations and changes. It therefore relies on two closely related areas: system change management and project management.

The main MSM goal is to help MS companies achieve excellence through the effective management of a continuous cycle of MSD projects. Therefore, managing necessary changes is one of the most important aspects of the framework. As an organization follows the continuous cycle of MSM, the level of changes involved depends on the scale of the MSD project, the amount of change required for the existing operations, and hence how the system structure is to be affected. The nature of change can be defined by three main variables:

- *Depth of change*: the degree to which the change affects the nature of the system—from incremental changes such as those normally associated with continuous improvement MSD projects, to profound changes such as those of greenfield or brownfield MSD types.
- *Speed of change*: the measure of the combination of depth and duration of the change. Although MSD projects are necessary, no MS organization can afford to spend too much time on their planning and execution.
- *Implementation of change*: how changes are introduced to the MS system. Change may be imposed, or may be the result of a total consensus. How it is introduced will have a significant impact on the company concerned.

It is necessary to make certain that the organizational changes are agreed, training needs are identified, and a training package is designed and provided. The ideal implementation teams should be multi-skilled and include team members from all affected departments, and from various positions in the organization. This will ensure commitment all over the company and increase the probability of success for both the system implementation and its future operation. Some of the most important aspects of change management are outlined in Figure 5.1, which are embedded in the relevant MSI task documents:

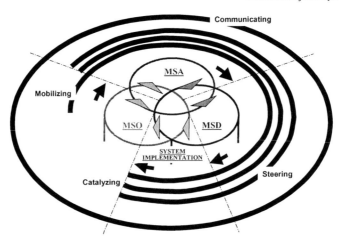

Figure 5.1 Change management within MSM framework

- *Mobilizing* initiates the actual process of change by making the organization mobile throughout the MSM change cycle. This consists of a sequence of unfreezing, transformation, and re-freezing. It draws attention to the actions of those involved in the change, and gives them reassurance that change is justified and that the project is being properly managed.
- *Catalyzing* deals with the creation of a structure that will enable and stimulate the implementation of change. Resources have to be made available and some have to be dedicated exclusively to it. The establishment and empowerment of *system improvement teams* is one of the means to help achieve this (*Worksheet OG.4*).
- *Steering* aims to keep the attitude of the interested parties, and the process of change themselves, on the right track. It predicts discrepancies between objectives and actual achievement, and then tries to use resources effectively. It should resolve any difficulties that arise and spread patterns of behavior that reinforce change.
- *Communicating* the vision of change to the employees at all levels of the organization hierarchy is vital. Initially, a high level of awareness of the strategic initiatives and objectives of the necessary changes should be maintained. It is then necessary to provide information on the progress of change, and to reassure all the affected parties outside the business.

5.2 PROCESSES OF IMPLEMENTATION

The general principles and techniques of change management presented above provide the basis for the actions of MS system implementation. The importance of implementation is easily seen because a decision or an intended system will not be of much use until properly implemented and effectively put to operation. Many cases have shown that difficulties associated with the implementation stage are the major obstacle to fully utilizing the potential benefit of the intended systems.

The entirety of the decision block, therefore, consists of two actions: making a choice and then implementing the change associated with that choice. The level of difficulty associated with implementation depends on the amount of changes required for the existing system, and how its structure and operation are to be affected. A carefully thought-out strategy will normally be required to carry out this last phase of an MSD project. The aim is to link the new system design, developed during the MSD phase, into transition plans and implementation programs which will lay a foundation for a successful implementation of the new system. Again, the three main aspects that are incorporated in the implementation phase are processes, IT, organization and human resources.

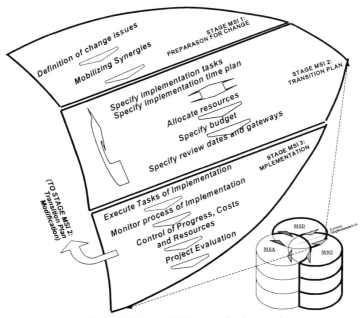

Figure 5.2 Stages of MS system implementation

As can be seen in Figure 5.2, this phase takes the outputs of the MSD phase as inputs, and begins with an assessment of the readiness for implementation within the organization. These results, together with the overall implementation plan of *Worksheet MSA/MSD 2.4.1*, provide the basis for detailed transition plan(s) to be specified. These plans will include scheduling, budgeting and resource requirements to bring the design of the new manufacturing/logistic systems up to date. Finally, the implementation stage actually makes the design a reality. This is achieved through the control of system installation by monitoring time, costs and the establishment. Again, project management software tools are highly recommended to help the planning, monitoring and management of MS system implementation.

Task Document MSI 1—Preparation for Change

TASK OVERVIEW

TASK DESCRIPTION

This task aims to make sure that the system—particularly the personnel in the organization who will be affected by the project—is ready for the changes required. Also, the MSD team and the system user should be given a common understanding of all the definitions used in the design, which is particularly important when the implementation of a distributed MS system is involved. The aim of the task is to:

- Confirm vision so that the strategic vision and the change actions are understood by all concerned.

- Convey improvement requirements so that those concerned are motivated by the evidence of the opportunities available.

- Gain employee participation through communication and the creation of a feeling of security that it is possible to achieve the improvements identified and evaluated in previous stages.

This task consists of the following activities:

- *Definition of issues at stake.* It is necessary to have a complete idea of the consequences of change for the various parties affected (e.g., employees, customers, suppliers, distributors, shareholders). When the consequences for each party are analyzed, the key stakeholders should be involved in the process from an early stage.

- *Mobilizing and steering synergies.* This aims to ensure that the need, urgency and purpose of change are understood and identified by the majority of those affected by the change, and to convert potential opponents into supporters. The techniques of influence analysis provide a systematic identification of key stakeholders and appraisal of their influence on, and attitude towards, the change. It may also involve creating a strategy to reshape the influence of these affected parties. In general, the key is to be able to implement change with and through people. Once the main issues have been identified, mobilization can be facilitated through: (a) seminars for groups of people consisting of those key stakeholders who potentially have the ability and power to make the change work; (b) workshops to initiate a dynamic of change at the operational level, to ensure that project or action is conceived from the operational staff, so that the workforce will be more committed.

TASK LINKS   *POSITION IN MSM FRAMEWORK*

INPUT FROM: MSA/MSD 2.4.1 (overview of implementation). New system design information from all MSD worksheets.

OUTPUT TO:

OUTPUTS Understanding and participation of workforce and others affected.

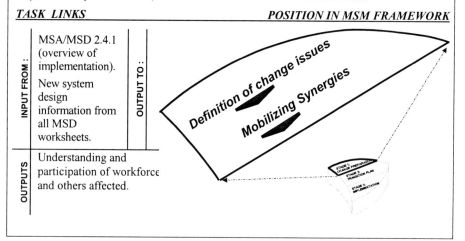

TASK PROCEDURE   **TASK FLOWCHART**

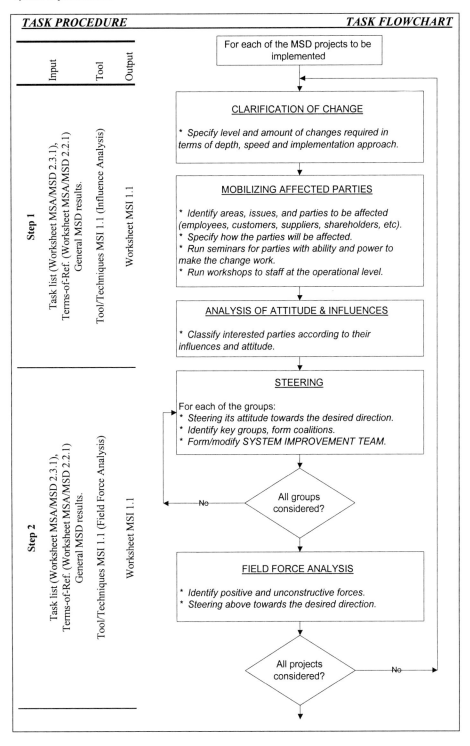

TOOL/TECHNIQUE MSI 1.1—Influence and Field-Force Analysis

The techniques of influence analysis can be used at any phase of the change process to gain the maximum amount of possible support for the project. This is done through the following steps:

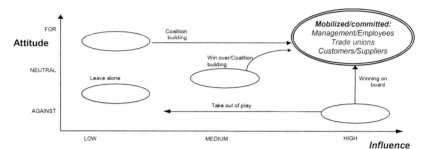

- Identify the key stakeholders.
- Evaluate their influence on the issue in question (high, medium or low).
- Evaluate whether they are currently for the change, against it, or neutral.

The results can be summarized in a diagram as shown. The situation may be improved by (a) bringing new key personnel into play, (b) boosting the influence of personnel who are currently in favor of the change, (c) reducing the influence of hostile personnel, (d) modifying the change content itself to secure more support. In relation to the above, another technique known as *field-force analysis* can also be used to evaluate the influences that have positive impact on the change process. This technique involves the following steps:

- Define the major factors of change according to the requirement of the project, in terms of: tangible forces, intangible forces, internal and external forces.
- Evaluate the impact from each of the above, scoring each force on a scale from 0 (low) to 10 (high).
- Determine actions based on changing the balance of forces.

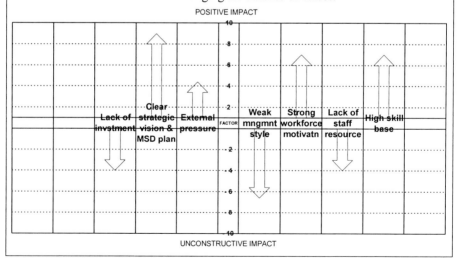

WORKSHEET MSI 1.1—Influence and Field-Force Analysis

Project Title:	**MSD Project title**:
Version:	**Date Completed:**

Influence Analysis Diagram

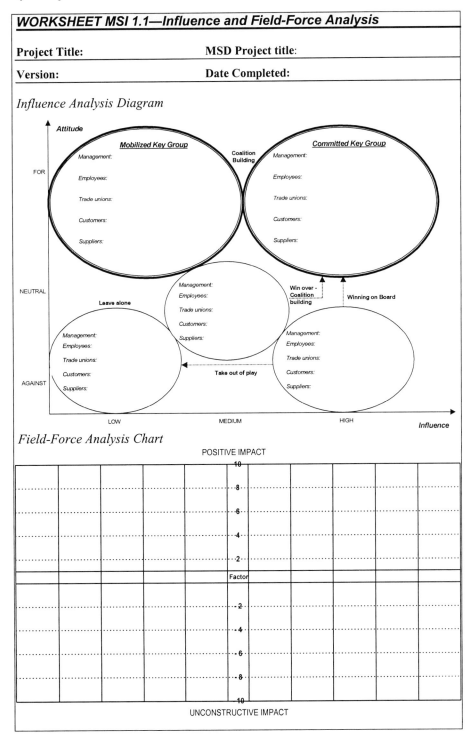

Field-Force Analysis Chart

Task Document MSI 2—Development of Transition Plan

TASK OVERVIEW

TASK DESCRIPTION

This phase aims to develop a transition plan for implementation, which should include a time plan, a resource allocation plan, budget, performance measures and contingency plan. It is important that the objectives of the changes have been defined and that each department in the organization fully consider the requirements of the new system. Employee skill and expertise must be exploited in order that the following are assured: (1) all the departments and parts of the business affected allocate the appropriate resources and the necessary time to the change tasks; (2) relevant skills and methodological support are brought into the project team; and (3) measures involved in the process are coordinated.

Accordingly, a coherent set of detailed plans and instructions must be prepared to guide the necessary actions to be taken, including the following items: outline of the requirement, description of method of implementation, specification of tasks, specification of personal requirements, allocation of resources for the tasks, and a time plan. Budgeting and election of performance measures are two closely related tasks. The budget is derived by estimating the cost of activities and resources. In general, the following steps are involved:

- Identify availability of resources and money.
- Check if the time plan is feasible, especially with respect to available resources.
- Identify points of no return (i.e., dates after which the investment will have been committed and the project will have to follow through) and include these in the schedule.
- Define dates of review points for monitoring and control processes, and important gateways which represent major milestones.

In addition, there are two types of performance measures that must be defined: (1) Performance measures to monitor and control the progress of the implementation. An example of such a performance measure would include comparing actual and budgeted cash flow for the project; (2) Performance measures to control the success of the new system itself. An example of such a performance measure would include customer service levels from the resultant system.

The output here is a detailed plan for implementation, which forms the basis for monitoring and controlling the progress of the project.

TASK LINKS   *POSITION IN MSM FRAMEWORK*

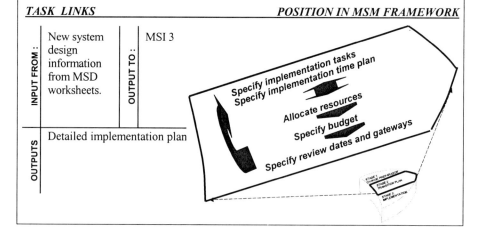

INPUT FROM:	OUTPUT TO:
New system design information from MSD worksheets.	MSI 3

OUTPUTS
Detailed implementation plan

TASK PROCEDURE   **TASK FLOWCHART**

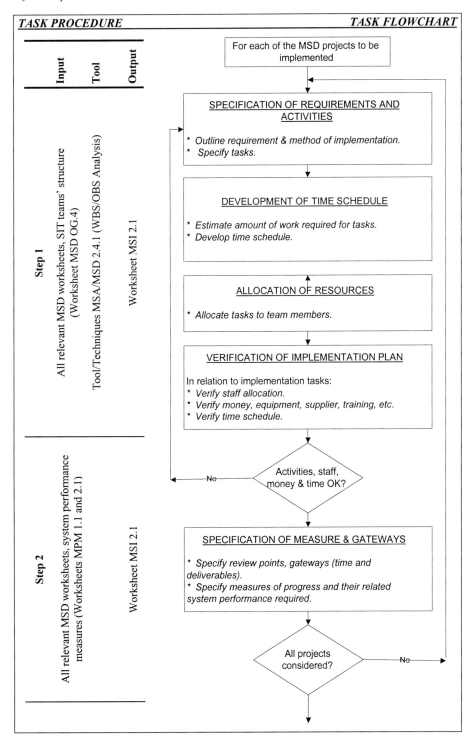

WORKSHEET MSI 2.1—Implementation Plan

System Project Title		MSD Project Title			Version		Date
System Owner		SIT Member					

	Implementation Activity Description	ID	Activity Duration (Weeks)
			1 2 3 4 5 6 7 8 9 10 11 12 13 14 15 16 17 18 19 20 21 22 23 24 25

REQUIREMENTS
- System Function
- System Structure
- System Decisions

CONCEPT'L DESIGN
- MS Processes
- Human & Organization
- Information & Control

DETAILED DESIGN
- Process
- Facilities
- Support
- Planning
- Control
- Human
- Organization
- Warehouse & Transport

Gateways/Review Points

Task Document MSI 3—MS System Implementation

TASK OVERVIEW

TASK DESCRIPTION

Having established the logical structure of the processes of implementation, in terms of time, resources and scope, this stage is concerned with the actual execution of the transition program, and monitoring and controlling the progress. These should be done in such a way that the project goal is achieved in the shortest possible time and at minimal cost. Steering is again a relevant issue here, requiring the following considerations during the stage: (1) Facilitating and accelerating the process of implementation by making sure it runs properly on a day to day basis; (2) Monitoring the attitude to change of key staff within the task team and providing advice and suggestions; (3) Identifying and making available useful tools, methods, training and coaching.

Once started the progress of implementation tasks needs to be continuously monitored to see that the changes are indeed taking place according to plan and if necessary feedback actions should be taken to adjust the individual tasks or even the course of the whole plan. Once implemented, the results of the changes brought about by the chosen option should be measured. These should be compared with the predicted outcome. Therefore, the overall process of systems implementation has the inherent characteristics of a feedback structure (execute, monitor, control) consisting of the following activities:

Execute. The execution of implementation consists of all the actions from start-up to go-live. The following are approaches that can be explored in practice: (1) Parallel running and going live. This approach has two purposes: to ensure that the transition from the existing system to the new system is achieved with the minimum amount of disruption, and to ensure the compatibility with other parts of the existing systems. With this approach, the new system or the new parts of the system can be run on their own for several cycles, while checks on compatibility can be carried out. (2) Old system shutdown. This shuts the old system down and opens the new one simultaneously. A decision is normally required to determine which parts/products/processes from the old system need to be retained and for how long, and what documentation needs to be created from the old system.

Monitor. The progress of implementation should be monitored and controlled. Proper control of the project progress depends on up-to-date status of the project. Therefore, information regarding project progress must be collected and analyzed on a continual basis. Such information helps the team identify potential problems if the implementation is not going as planned. Progress of system implementation may be monitored in terms of:

- Task progress: the work completed on a task to date, as against the planned start and finish dates.
- Costs: how much a particular resource costs on a certain phase, or how much total cost has accrued.
- Resource utility: work done by a resource as against the work the resource is scheduled to undertake in a particular phase. These parameters can be measured in terms of the actual amount/work done to date, or as a percentage of the scheduled amount.

Control. In reality, every project has variances. It is important that when deviations in the operations are detected, steps must be taken to counteract these and the implementation program must be altered to put the project back on course.

TASK DESCRIPTION

Control should be exercised in regards to task progress, costs, and resources. However, it should be realized that adjustment of one of these may necessitate readjustments of others:

- *Progress.* To keep the overall project on schedule, teams must make sure that the individual tasks start and finish on time. It is important to identify tasks that vary from the original transition plan as early as possible so task dependencies and resources can be rescheduled to meet the next deadlines.
- *Costs.* Decisions need to be made, whenever it is detected that the available cash is exceeded, whether changes are necessary to finish the project within budget.
- *Resources.* The utility of people and equipment needs to be checked to see whether the resources are located with too much or too little work. These need to be balanced by: (a) Adding more resources to a critical task, (b) reassigning a task/reminder of a task to another resource, (c) delaying non-critical tasks assigned to an overworked resource.

In practice, commercially available project management software is highly recommended for the monitoring and control of an MSD project. Also, to manage a project effectively, the team or teams need to communicate project information promptly. In this respect, an added benefit provided by some software includes tools that can be used to help manage a distributed MSD project. For instance, Microsoft Project helps set up workgroups located in different geographical locations, providing a means of electronically linking team members and making it easier for them to exchange information about the project. Once such a setup is complete, information about a project can be communicated, collected and distributed through the following:

- *The Web* allows collaborative planning among workgroup members, project managers, and other stakeholders by providing access to project details. Workgroup members can view and manipulate details of their assigned tasks, and check the latest information for the entire project. Workgroup members can also create new tasks and send them for incorporation into the overall project file, as well as delegate tasks to other workgroup members. Project managers can request, receive, and consolidate status reports from members either on- or off- site.
- *E-mail* allows a workgroup to be connected, and facilitates assigning tasks, requesting and submitting status reports, and sending and receiving task updates.

TASK LINKS   *POSITION IN MSM FRAMEWORK*

INPUT FROM :
Worksheet
MSI 2.1
(Transition plan).
System design
data from MSD
worksheets.

OUTPUT TO :
Wksheet
MSI 3.

(TO STAGE 2:
Transition Plan
Modification)

Execute Tasks of Implementation
Monitor process of Implementation
Control of Progress, Costs
and Resources
Project Evaluation

OUTPUTS
Modified transition plan.
Implementation of new system.

TASK PROCEDURE   TASK FLOWCHART

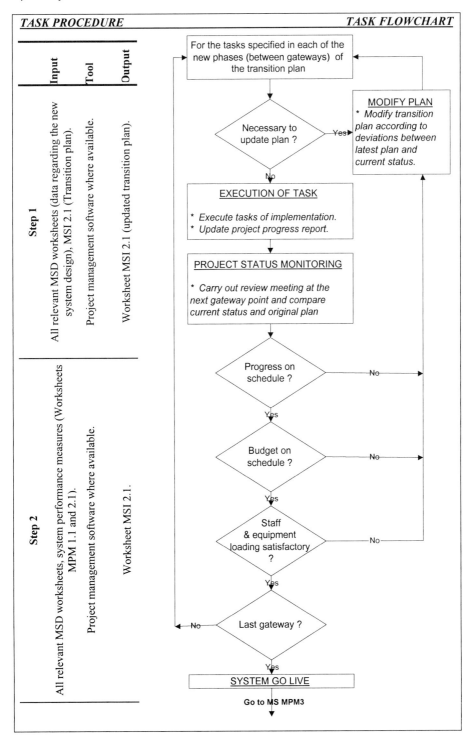

Input | **Tool** | **Output**

Step 1

All relevant MSD worksheets (data regarding the new system design), MSI 2.1 (Transition plan).

Project management software where available.

Worksheet MSI 2.1 (updated transition plan).

Step 2

All relevant MSD worksheets, system performance measures (Worksheets MPM 1.1 and 2.1).

Project management software where available.

Worksheet MSI 2.1.

For the tasks specified in each of the new phases (between gateways) of the transition plan

Necessary to update plan ? — Yes→

MODIFY PLAN
* *Modify transition plan according to deviations between latest plan and current status.*

No

EXECUTION OF TASK
* *Execute tasks of implementation.*
* *Update project progress report.*

PROJECT STATUS MONITORING
* *Carry out review meeting at the next gateway point and compare current status and original plan*

Progress on schedule ? —No→

Yes

Budget on schedule ? —No→

Yes

Staff & equipment loading satisfactory ? —No→

Yes

←No— Last gateway ?

Yes

SYSTEM GO LIVE

Go to MS MPM3

CHAPTER SIX

MS Performance Measurement and System Status Monitoring

6.1 INTRODUCTION

As pointed out in Chapter 1, an MS organization's performance measurement should be an integral part of its MSM framework, and should play a vital role in directly supporting the achievement of the organization's strategic goals. The objectives and goals of the organization should be clearly in line with the system purpose, as specified by its upper system and environment. This applies at each level of the organizational tree. That is, the strategies and policies adopted at various levels within the organization must be coherent and in harmony with the overall organizational objectives, as shown in Figure 6.1. The ability to develop and achieve such a set of coherent strategies and aims must be regarded as one of the key issues of MS systems management.

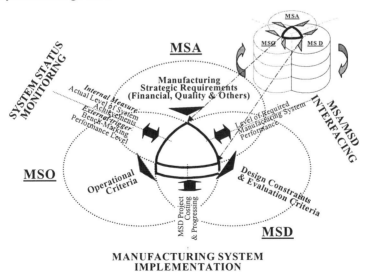

Figure 6.1 MS performance measures

The continuing awareness of what is happening in the wider business environment is another prerequisite for the system's effective operation. Sufficient consideration must continually be given to the influence of environmental factors such as the change of government policies and institutional regulations, economical and political climate, as well as customer requirements and technological development. This explains why benchmarking is important to the success of an MS organization. Therefore, in accordance with these pre-conditions for efficient systems operation, MSM performance measurement and system status monitoring are needed to:

- clarify customer requirements,
- help understand the progress of business processes,
- ensure decisions are based on fact, not on emotion, and
- show continuously where improvement needs to be made.

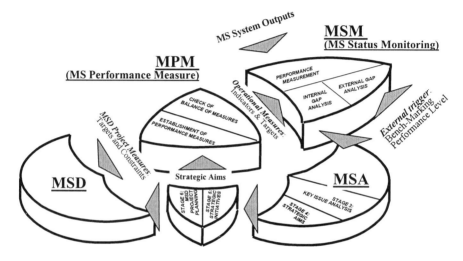

Figure 6.2 Stages of system performance monitoring within MSM

Therefore, performance measure setting and system status monitoring together form an integrated cycle, providing a tool to check consistency between strategic objectives and performance measurement. Since it is based upon a system's perspective of company performance requirement, the cycle prevents local optimization by combining more than one aspect of performance within the overall MSM framework, and throughout the complete MSA/MSD/MSO cycle. As an integral function, it can also help a company focus on improving the competitiveness of its MS system as a whole, and on motivating continuous improvement. By closing the MSA-MSD-MSO loop, this cycle helps to accomplish an overall control of the manufacturing/supply system. Such a self-regulation mechanism provides the ability to continuously adapt to the environmental changes, and is one of the prerequisites for the survival of open systems like MS organizations.

The overall structure of the MSM performance monitoring module is as shown in Figure 6.2, consisting of two stages: MS performance measure (MPM) and MS status monitoring (MSM). As can be seen, performance monitoring is closely related to the MSA process, with a certain degree of overlapping between the two. The reason for this is obvious: in order to ensure that an MS system achieves a strategically competitive position and that different parts of the organization are pulling their weight in a combined effort to maintain this position, some form of coherent performance monitoring of both individual units as well as the whole is essential. The ultimate aim of performance measurement is to motivate behavior leading to continuous system improvement. This can only be achieved by evaluating and quantifying the current state of the company, and highlighting where progress has been made and which areas need to be improved. By using performance measures that support a company's strategy, the feedback from the process will provide the company with the information needed for ongoing improvement. This allows for monitoring the critical success areas so that corrective actions can be taken should a drift occur. Therefore, this module will assist in monitoring and initiating the right action whenever necessary in the manufacturing and supply process:

- *Specification of strategy-oriented performance measures.* The purpose of this is to disaggregate strategic requirements into operational level criteria, and then measure the current system according to the relevant parameters.
- *Overall system status monitoring.* Based on the operational level measurements of the current system, this section produces an integrated assessment of the system's overall performance against its current strategic goal. It also determines whether further actions are needed and, if so, identifies the necessary programs of continuous improvement.
- *Continuous improvement monitoring.* The purpose and structure of this section is similar to the above. However, the focus here is the monitoring and assessment of the improvement of system performance as a direct result of the MSD actions initiated.

The task documents and their worksheets provide a method of assessing performance measures and analyzing system status.

6.2 MPM—SPECIFICATION OF STRATEGY-ORIENTED MEASURES

In practice, performance information should be used at all levels of management to drive performance improvement. It tells the management of an MS system about its present condition, and allows management to objectively measure the current system performance against others through benchmarking. In turn, benchmarking aids in identifying potential areas of performance improvement and in generating innovative ideas to drive that improvement. Specifically, the performance information here can be used to initiate two main types of MSD projects: the design of new MS systems, or future improvement of an existing operation.

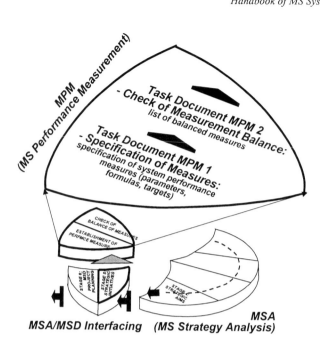

MSA/MSD Interfacing   **(MS Strategy Analysis)**

Figure 6.3 MPM—Specification of strategy-oriented performance measures

It is important that the performance measurement of the monitoring function be based on the clear identification of the business processes that have the most impact on the success or failure of the organization's goals. Therefore, when designing performance measurement systems it is necessary to decide what to measure, and how to measure it. As an integral part of the MSM system, the performance measures should always be tied to the system's current goals or objectives. Thus, a performance measurement system enables the organization to ensure it is progressing in the right direction as it moves from its current state to a future state along its system life cycle. Within the MSM framework, the information summarized in the MSA worksheets will provide the basis for the system status monitoring function. With such a foundation to provide the direction and reason, quantitative objectives can be defined to assess progress toward the vision. As a core area of the MSM, therefore, the MPM area aims to specify a set of strategy-oriented performance measures for the other functional areas within the framework. As shown in Figure 6.3. It consists of two task documents, aiming to help align performance measures with the previously established MS strategy. The following are the key points in this stage:

- measure only what is important,
- focus on customer needs,
- involve employees in the choice and implementation of the measures.

Task Document MPM 1—Specification of Measures
TASK OVERVIEW

DESCRIPTION

There are a number of sources that should be examined as a first step in establishing a set of meaningful and integrated performance measures:

- Outputs of the strategic planning process specify the company's mission and what directions the company should move to achieve those objectives.
- Analysis of key processes and factors having the most impact on the success or failure of the organization's goals.

The main inputs to this stage are the strategic initiatives from *Task Document MSA/MSD 1.1 (Worksheet MSA/MSD 1.1.1)*. The only additional requirement is the need for a clear definition of the key processes for the strategic aims. Following this, the task aims to establish a mechanism of performance measurement that directly supports the previously specified manufacturing/supply strategic requirements. This is achieved through the following steps:

- For each of the key processes listed in *Worksheet MSA/MSD 1.1.1*, identify its related key success factors, and set performance goals that are to be achieved at the end of the monitoring period.
- Specify parameters/measures of performance.
- Set overall performance indicators and achievement levels or targets. To help this key process, a linking-table is provided (*Tool/Technique MPM 1.1*) to illustrate the generic relationships between performance parameters and performance indicators. This cause-effect table can help identify the correct indicators to use for strategic concerns.
- Identify relevant matrices/tools such as formulas, utility weightings for the individual measures, and algorithms for calculations purposes.

TASK LINKS   POSITION IN MSM FRAMEWORK

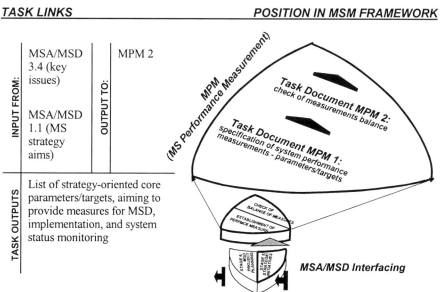

INPUT FROM:		OUTPUT TO:
MSA/MSD 3.4 (key issues)		MPM 2
MSA/MSD 1.1 (MS strategy aims)		

TASK OUTPUTS

List of strategy-oriented core parameters/targets, aiming to provide measures for MSD, implementation, and system status monitoring

MSA/MSD Interfacing

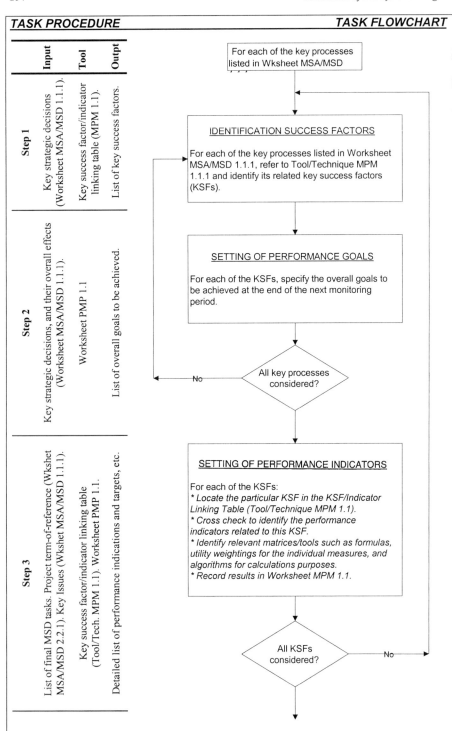

TOOL/TECHNIQUE MPM 1.1—Key Success Factor/Performance Indicator Link

Performance Measures		Quality																															Delivery/Lead time															
	Performance Indicators →	Rework rate	Safety	Scrap rate	Pass rate	Field failure	Reject rate	Material yield	Packaging quality	Supplier quality	No of complaints	Warranty claims	Service call rate	Supplier certification	Process capability	Expected product life	% conform to targets	% with no repair work	Product performance	Meeting project milestones	Assembly line defects rate	Lapse, renewal, retention rate	Quality relative to competitors	Hiring and layoff rates	Mean time between failures	% inspection operations eliminated	Perceived relative quality performance	Product reliability relative to competitors	Product durability relative to competitors	Conformance to specifications	No of engineering change	Hours of training per worker/year	Lead times through department	Response time	% on time for rush jobs	Paperwork throughput time	Material throughput time	Value added as % of total elapsed time	Decision cycle time	Time lost waiting for decision	Breakeven time	Hours of overtime	Time from idea to market	Average time between innovations	No of changes in projects	Non-manufacturing lead time	Average age and education of workers	Ratio of non-value added to value added activities
Quality	Incoming quality	X		X	X		X			X				X	X			X	X		X	X					X										X											
	First pass yield		X	X			X	X		X											X	X				X																					X	
	Not right first time			X			X																																									
	Process waste yield																								X					X																	X	
	Reject rate					X																									X																	
	Customer return rate				X						X	X	X				X											X	X	X																		
	Customer satisfaction	X									X	X	X				X		X								X	X	X	X								X										X

Performance Indicators / Performance Measures matrix

Performance Measures (columns):

Quality — 1 Incoming quality, 2 First pass yield, 3 Not right first time, 4 Process waste yield, 5 Reject rate, 6 Customer return rate, 7 Customer satisfaction
Delivery Lead time — 8 Order process times, 9 Time between order and delivery, 10 Vendor lead times, 11 Manufacturing cycle time, 12 No. of change in project, 13 Manufacturing lead time
Delivery Reliability — 14 % on time deliveries, 15 Delivery schedule achievement, 16 Inventory accuracy, 17 Forecast accuracy, 18 No of days of late shipments, 19 No of complaints of shipping damage
Volume Flexibility — 20 Ability to response to demand increase, 21 Lot size, 22 Worker flexibility, 23 Capacity inbalance, 24 Smallest economic volume
Design Flexibility — 25 Proportion customerized, 26 Variety flexibility, 27 Labour skill, 28 Design change per year, 29 Ability to introduce new product, 30 Ability to cope with product change
Cost — 31 Unit manufacturing cost, 32 Overhead cost, 33 Direct labour productivity, 34 Indirect productivity, 35 Raw materials inventory turnover, 36 Finish goods inventory turnover, 37 Absenteeism

Performance Indicators (rows) with marked (X) measures:

Group	Performance Indicator	Marked Performance Measure(s)
Delivery reliability	Speed of set ups	7
	Lead times for raw materials	8, 9
	Perceived relative reliability	1, 9
	Reliability relative to competitors	9
	% of orders with incorrect amount	9
	Schedule attainment	1, 8, 9, 11
	Due date adherence	13
	Average age of equipment	23
	% improvement in output	12
	Forecast accuracy	13
	Overall equipment effectiveness	17
	No of suppliers	10, 17
	Variety flexibility	26
Volume Flexibility	% reduction in lead time per product line	4
	Set-up time	11
	Lot size	20
	Job classification	27
	Vendor lead time	9
	Capacity utilisation	29, 30
	% of slack time for labour	27
	% multipurpose equipment	23
	% of programmable equipment	20
	% increase in direct labour skills	27, 29
	Product cycle time	29, 30
	Volume flexibility relative to competitors	20
	Process flexibility relative to competitors	29
	Time to replace tools, change tools	11
	Perceived relative volume flexibility	20
	How well plant adapts to volume change	20, 24
	Smallest economical volume	22, 24
	Workforce cross-trained	27
	Schedule change ability	30
	Ability to response to demand increase	20
	Strategic control system	37
	Product shelf life	34
	Minimum/Maximum order size	20, 22
	Seasonal demand variation	23
	Random demand variation	23
	Frequency of schedule change	24, 28
	Size of schedule change	28
	Effect on delivery lead-time	9, 11
	Ability to perform multiple tasks efficiently	22

Performance Indicators (columns) × Performance Measures (rows)

Indicator groups: columns 1–17 = **Design Flexibility**; columns 18–51 = **Cost**

Measure group	Performance Measure	Skill level	Perceived design flexibility	Design flexibility relative to competitors	Perceived relative product flexibility	No of material parts	Speed of new product introduction	Speed of new model introduction	Ability to cope with product range	No of product types produced	No and type of engineering change orders	Ability to cope with product change	Design change per year	Introduction of new processing equipment	Customisation ability	speed of decision making to respond to market needs	% increases in lead-time over standard product	Development time for new products	Cost of quality	Tooling cost	Overtime cost	Repair/re-work cost	Manufacturing cost	Material cost	Energy/Utility cost	Cycle time cost	Downtime cost	Absenteeism cost	Stock/Debtors cost	Overhead cost	Cost relative to competitors	Perceived relative cost performance	Capital productivity	Machine productivity	Direct labour cost	Indirect labour cost	% improvement in labour	Relative labour cost	Labour efficiency	Inventory cost	Scrap cost	Design cost	Relative R&D expenditure	Distribution cost	Funds for acquiring new technology	Maintenance expenses	Funds for training programs	Life cycle cost	Service cost	Cost of non-value added activities	Dollars of capital investment	Warranty cost
Quality	Incoming quality																																																			
	First pass yield																																																			
	Not right first time																																																			
	Process waste yield																																																			
	Reject rate																																																			
	Customer return rate																																																			
	Customer satisfaction																																																			
Delivery — Lead time	Order process times																																																			
	Time between order and delivery																																																			
	Vendor lead times																																																			
	Manufacturing cycle time																									x																										
	No. of change in project																																																			
	Manufacturing lead time																																																			
Delivery — Reliability	% on time deliveries																																																			
	Delivery schedule achievement																	x																																		
	Inventory accuracy																																																			
	Forecast accuracy																																																			
	No of days of late shipments															x	x																																			
	No of complaints of shipping damage																																																			
Volume Flexibility	Ability to response to demand increase													x	x																																					
	Lot size																																																			
	Worker flexibility	x																																																		
	Capacity inbalance																																																			
	Smallest economic volume																																																			
Design flexibility	Proportion customerized		x		x	x																																														
	Variety flexibility								x	x																																										
	Labour skill	x																																																		
	Design change per year			x									x																																							
	Ability to introduce new product						x	x				x																																								
	Ability to cope with product change										x	x																																								
Cost	Unit manufacturing cost																		x	x		x	x	x																												
	Overhead cost																		x	x	x	x			x	x	x			x	x	x	x	x	x						x	x	x	x	x	x	x	x	x	x	x	x
	Direct labour productivity																																		x		x	x	x													x
	Indirect productivity																																			x	x	x	x													
	Raw materials inventory turnover																												x											x												
	Finish goods inventory turnover																												x											x												
	Absenteeism																											x																								

WORKSHEET MPM 1.1—*Definition of Performance Measures*

Project Title:

Person(s) Responsible:

Version: **Date Completed:**

Key decision areas	Key success factors (measures)	Performance indicators (parameter/target)					
		Quality	Delivery lead-time	Delivery reliability	Design flexibility	Volume flexibility	Cost
1							
2							
3							
4							
5							
6							
7							

Task Document MPM 2—Balance of Measures

TASK OVERVIEW

DESCRIPTION

Having identified the relevant measures, this task is designed to review the measures and make necessary adjustments so that they stimulate purposeful action when put into use. The aim is to balance internal and external requirements, as well as financial and non-financial measures. As a general guide to the principles of balanced measures, it should be pointed out that the approach adopted here is in contrast to traditional performance measures which have been primarily based on management accounting systems. Traditionally, performance measures have been confined to cost-related performance measures, focusing on financial data such as profit, return on investment or cash flow. If only used properly, they tend to produce localized optimization of individual units. In adopting a more balanced view, this task uses a mix of measures for system status/performance monitoring. This is achieved by answering four basic questions:

- From the customers' perspective: how do the measures affect the customers' view on company performance?
- From an internal perspective: in relation to the strategic aims/initiatives, what measures help the company to achieve what it must excel at?
- From the perspective of innovation and learning: what measures help the company continue to improve and create value?
- From the financial perspective: how do the measures affect the shareholders' view on the company performance?

For each of the above, goals should be set by identifying the specific measures and targets, so that the contents of *Worksheet MPM 1.1* can be checked, balanced and finalized. The techniques used to deal with conflicting objectives (*Task Document MSA/MSD 2.2*) apply equally here, should this become an issue.

Financial Perspective	
Goals	**Measures**
Survive	Cash flow
Succeed	Quarterly sales growth and operating
Prosper	income by division
	Increased market share

Customer Perspective	
Goals	**Measures**
New products	% of sales from new products
Responsive supply	On-time delivery
Preferred supplier	Share of key accounts' purchases
Customer partnership	Number of cooperative engineering efforts

Internal Business Perspective	
Goals	**Measures**
Technology capability	MS geometry vs. competition
Manufacturing excellence	Cycle time
	Unit cost yield
Design productivity	Engineering efficiency
Product introduction	Actual introduction schedule vs. plan

Innovation and Learning Perspective	
Goals	**Measures**
Technology leadership	Time to develop next generation
MS learning	Process time to maturity
Product focus	Percent of products that equal 89% sales
Time to market	New product introduction vs. competition

An example of a balanced scorecard

TASK LINKS

INPUT FROM:	MPM 1.	**OUTPUT TO:** All relevant MSM functional areas as shown in Fig 10.1.

TASK OUTPUTS

Specification of strategy-oriented, balance-checked core parameters/targets to provide measures for MSD, implementation, and system status monitoring.

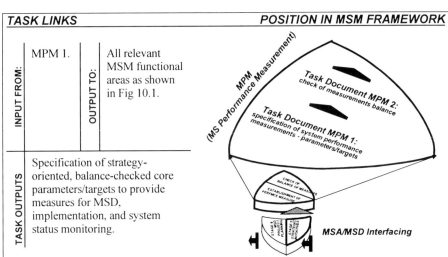

MSA/MSD Interfacing

TASK PROCEDURE

TASK FLOWCHART

	Input	Tool	Output
Step 1	List of performance indicator.	Worksheet MPM 2.1.	Indicators listed under the categories in Worksheet MPM 2.1.
Step 2	Results from previous step.	Worksheet MPM 2.1. Tool/Tech. MPM 1.1.	Balanced indicators listed under all of the four categories in Worksheet MPM 2.1.
Step 3	Results from previous step.	Technique used to deal with conflicting objectives (Task Document MSA/MSD 2.2).	Refined, balanced performance indicators, listed in Worksheet MPM 2.1.

For the list of previously specified performance indicators:

Identify and list:

* *indicators from customer perspective*
* *indicators from financial perspective*
* *indicators from internal perspective*
* *indicators from innovation/learning perspective*

Any missing category ? — No

Identify missing items to produce a balanced profile of indicators

Any conflicting indicators ? — No

Refine list of indicators to resolve conflicting ones

WORKSHEET MPM 2.1—Balance of Performance Measures

Project Title:

Person(s) Responsible:

Version: **Date Completed:**

Measurement *Internal* *(week/month)*	Performance Indicators (Parameter/Target)					
	Quality	Delivery Lead-time	Delivery Reliability	Design Flexibility	Volume Flexibility	Cost
Financial Perspective						
Customer Perspective						
Internal Business Perspective						
Innovation and Learning						

6.3 MSM—MS SYSTEM STATUS MONITORING

This section consists of a number of steps, as shown in Figure 6.4. The initial steps of this stage are a reversal of the previous stage. The previous stage links the overall strategic concerns to operational level parameters through a process of disaggregation. In contrast, having taken measurement of system performance based on the relevant parameters identified through the disaggregating process, this stage aggregates these values back to a higher level, allowing the systems performance to be assessed according to the original strategic goals. As was illustrated in Figure 6.2, there is significant overlap between the last two sections of the performance monitoring module and the MSA model. Consequently from this point on the process flow, the steps, concepts, techniques and considerations involved have a great deal in common with those of MSA. A detailed description and discussion are therefore not necessary.

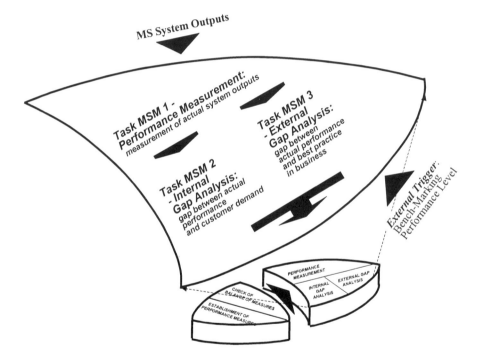

Figure 6.4 MSM—MS status monitoring

However, it is crucial here to distinguish between the internal and external system performance gaps. The difference between these is illustrated in Figure 6.4. Whereas the internal gap helps a company identify the difference between its market requirement and its current systems performance, the external gap is based on the current best-practice through benchmarking. Both provide an indication of future requirements.

Task Document MSM—MS Status Monitoring
TASK OVERVIEW

DESCRIPTION

This is the stage that monitors the performance achieved by a new or improved MS system resulting from current MSA initiatives and related MSD projects. The main aims are to evaluate the effectiveness of MSD activities with respect to meeting the strategic objectives, and to motivate further behavior leading to continuous improvement in customer satisfaction and productivity.

Having carried out the previous MSA/MSD cycle of analysis, the processes involved here are straightforward. This task is fundamentally a repeat of the MSA cycle, aiming to identify requirement/performance gaps and then, if necessary, to specify future initiatives, action plans and the relevant MSD actions. The only differences are:

- Measurement indicators used in this stage are more focused, using the list of company-specific measures already identified through the previous MSA cycle instead of the list of general measures used initially in *Worksheet MSA 2.3.1*.
- Benchmarking is regarded here as an integral part of the process. It can be used by companies to compare performance and to find and implement the best practices. It involves systematically and continually comparing the performance of an organization against the performance of the business leaders. It is a useful approach that can be adopted at this stage to develop the future actions needed to achieve winning strategies. This is done by identifying superior performance against others in the market, with the aim of achieving a world-class standard.

The stage therefore consists of the following steps:

- *Measure system performance as a result of the current action plans/MSD projects.* The objective of this is to identify action plan performance. Although it may be difficult to measure individual performance as a result of specific action plans, the overall product and system performance should generally reflect the effects of action plans previously executed.
- *System performance profiles.* Based on the performance measurements obtained through the previous step, this produces a number of system requirement/performance profiles. The purpose and approaches to be used are identical to those of the MSA process (*Stage MSA 2.3* and *2.4*).
- *Internal gap analysis.* This compares the current system performance with the targets set previously in *Stage MPM 1*, and identifies the requirement/ performance gap.
- *External gap analysis.* This compares the current system performance with that achieved by competitors, and identifies the differences between the company's current performance and the best practice.
- *Identify future actions necessary.* Within the context of MS system status monitoring, three possible courses of action can be taken depending on the results of the analysis:
  1) Current performance satisfactory—no need for immediate actions.
  2) Current performance unsatisfactory—identify/modify action plans for the current initiatives to improve current MSD project yields.
  3) Current performance unsatisfactory—initiate new MSA/MSD cycle.

TASK LINKS

INPUT FROM: MPM 1.

OUTPUT TO: Repeat MSM 1 if satisfactory. Initiate new MSA/MSD cycle otherwise.

OUTPUTS: Measurement of current system performance.
Current system status in terms of performance/requirement gaps (both internal and external).

POSITION IN MSM FRAMEWORK

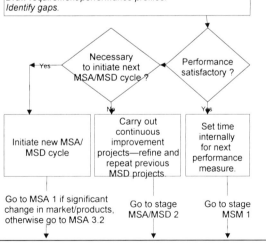

Task MSM 1 - Performance Measurement: measurement of actual system outputs
Task MSM 2 - Internal Gap Analysis: gap between actual performance and customer demand
Task MSM 3 - External Gap Analysis: gap between actual performance and best practice in business

TASK PROCEDURE

	Input	Tool	Output
Step 1	Performance indicators	Wksheet MSM 1	Performance measurement
Step 2	Results from Step 1	Wksheet MSM 2	Int. perf. gaps
Step 3	Profiles from last step	Benchmarking (Tool/Tech. MSM1)	Existing performance gaps
Step 4	Profiles from last step	Relevant tasks in the MSA area Worksheet PMP 1.1	Decisions regarding future MSM activities

TASK FLOWCHART

PERFORMANCE MEASURE

For the previously specified performance indicators as listed in Worksheet MPM 1.2.1:

Measure system performance.
Record results.
Estimated the aggregated overall performance against the competitive criteria.

INTERNAL GAP ANALYSIS

For each of the previously specified Product groups/system:

Draw requirement/performance profiles.
Identify gaps.

EXTERNAL GAP ANALYSIS

For each of the previously specified Product group/system:

Carry out benchmarking exercise and identify competitor and/or best-practice performance levels.
Draw requirement/performance profiles.
Identify gaps.

◆—Yes— Necessary to initiate next MSA/MSD cycle ?

Performance satisfactory ?

No

Yes

| Initiate new MSA/MSD cycle | Carry out continuous improvement projects—refine and repeat previous MSD projects. | Set time internally for next performance measure. |

Go to MSA 1 if significant change in market/products, otherwise go to MSA 3.2

Go to stage MSA/MSD 2

Go to stage MSM 1

TASK TOOL/TECHNIQUE MSM 1—Benchmarking

Although a number of benchmarking process models have been suggested in the literature, they share some common elements such as *plan*, *collect data*, *analyze* and *act*, which are in fact already embedded into the MSM framework.

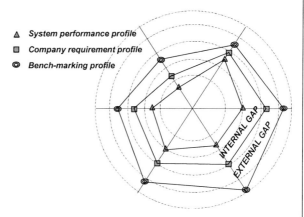

△ **System performance profile**

▣ **Company requirement profile**

◎ **Bench-marking profile**

The key to benchmarking lies within what to benchmark, and with what to set the benchmark. Three different types of benchmarking can be used:

- *Internal benchmarking*. This compares an organization's internal activities and processes with the objectives to establish standards within the organization.
- *Competitive benchmarking*. This involves the investigation of a direct competitor with the aim of identifying a company's current position compared to market or industry standards. The purpose of carrying out such an exercise is to enable companies to compare their performance with competitors in the same industry. In a competitive benchmarking situation, there is a need to find similar companies. The fact that there are always differences between companies needs to be taken into consideration. These differences may include product, technology, workforce, management process, etc. Therefore, as suggested within the performance monitoring module, it is important to utilize a number of performance measures to achieve useful comparisons between companies.
- *Generic benchmarking*. This is based on a comparison of unrelated industries, using the results from the discovery of a best practice that is not utilized in the benchmarking firm's own industry.

Potential benefits of benchmarking

Objective	Without benchmarking	With benchmarking
Defining customer requirements	Based on history, acting on perception	Based on market reality, acting on objective evaluation
Establishing effective goals	Lacking in external focus, reactive, lagging behind	Credible, customer-focused, proactive, leading industry
Developing performance measures	Strengths and weaknesses not understood	Solving real problems, performance outputs known, based on "best in class"
Becoming competitive	Internally focused, evolutionary change, low commitment	Understanding the competition, incorporating innovative ideas with proven performance, high commitment
Industry practices	Not invented here, few solutions, small step continuous improvement	Proactive search for change, many options, breakthroughs

The most important benefit of benchmarking is that it allows a company to see beyond its existing paradigms of process performance. As it benchmarks other organizations, it greatly improves the likelihood of seeing tomorrow's solutions to today's problems, and of adopting a wider reaching strategy. The potential benefits of benchmarking are summarized in the table above. According to the process level involved within the organization, benchmarking can also be divided into the following classes:

- *Process benchmarking* focuses on work processes or operating units to produce bottom line results, such as increased productivity, reduced cycle time, lower costs, improved sales, reduced error rates, and improved profit.
- *Performance benchmarking* focuses on product and service comparisons such as price, technical quality, or service comparisons and analysis of operating statistics.
- *Strategic benchmarking* examines how companies compete. A key objective is to identify the winning strategies of highly successful companies.

In discussing performance monitoring, it is also necessary to talk about self-assessment because performance measurement is closely related to total quality management (TQM). An exercise of self-assessment involves selecting an appropriate model for comparison, collating data from suitable sources and having a comparison and scoring process that compares the two. In carrying out such an assessment, one first looks at existing models and chooses a model of excellence against which to assess one's own organization. Typically, this involves the self-assessment processes of a proven excellence measure , which can be used as a basis for development of a company-specific process. Such assessment is normally comprehensive and systematic, with the advantages of: being market led, being business- and internationally-backed, taking all essential elements into consideration, continuously evolving, representing the best practice for successful organizations, and whose core values fit most organizations.

Several national and international quality awards have been established to promote quality and serve as models of TQM. The most widely used frameworks include the European Quality Award and the Malcolm Baldrige National Quality Award (USA). For example, the Baldrige Quality Award for performance excellence and its scoring guidelines present a diagnostic instrument to help an organization identify organizational strengths, as well as key areas for improvement. The award also stresses the tacit characteristics of an organization such as leadership, commitment and involvement of employees. It consists of seven categories and a 1000-point scoring system. Its performance excellence criteria include a number of basic elements: leadership, strategic planning, customer and market focus, information and analysis, human resource development and management, process management and business results. Criteria are based on a number of core values, such as customer-driven quality, leadership, continuous improvement and learning, employee participation and development, fast response, design quality and prevention, long-range view of the future, management by fact, partnership development, company responsibility and citizenship, and results focus. The award is basically a measure of a company's competitiveness, and it places a great emphasis on continuous improvement in response to market pressure from customer demands, competitors and acceptable industry standards and performance. Companies participating in the award process are required to submit application packages that include responses to the award criteria. Award recipients are expected to share information about their successful performance strategies with other organizations. More details of these awards can be found in their respective web-sites.

WORKSHEET MSM 1—*Current System Performance*

Project Title:

Person(s) Responsible:

Version: **Date Completed:**

		A	B	C	D	E	F	System
					Product group			
Quality Indicator	1. 2. 3. 4. 5. *Overall*							
Delivery lead-time	1. 2. 3. 4. 5. *Overall*							
Delivery reliability	1. 2. 3. 4. 5. *Overall*							
Design flexibility	1. 2. 3. 4. 5. *Overall*							
Volume flexibility	1. 2. 3. 4. 5. *Overall*							
Cost/Price Indicator	1. 2. 3. 4. 5. *Overall*							
Other criteria								

WORKSHEET MSM 2—Summary of System Performance

Project Title:

Person(s) Responsible:

Version: **Date Completed:**

Performance Measurement

Product group		Quality	Delivery lead-time	Delivery reliability	Design flexibility	Volume flexibility	Cost/ price
A	Current target Current performance Best performance *Internal gap* *External gap*						
B	Current target Current performance Best performance *Internal gap* *External gap*						
C	Current target Current performance Best performance *Internal gap* *External gap*						
D	Current target Current performance Best performance *Internal gap* *External gap*						
E	Current target Current performance Best performance *Internal gap* *External gap*						
F	Current target Current performance Best performance *Internal gap* *External gap*						
System	Current target Current performance Best performance *Internal gap* *External gap*						

CHAPTER SEVEN

Institutionalization of MSM —Application and Tools

7.1 INTRODUCTION

In order to fully utilize the potential of the MSM framework, its concepts should be integrated into a company's management system and culture, and its procedures should be institutionalized as an integral part of the organization's operation. The following are some of the key considerations:

- *Competent people*. A prerequisite is to have managers and staff with the right attitude, motivation, skills and training, who know why they are designing or reengineering an MS system, and how to do it.
- *Competent organizational structure*. The organization needs to be set-up to support the necessary MSM activities. This should provide a structure and means for monitoring the current system status, analyzing its operational and strategic needs, and accordingly, initiating and authorizing MSM projects. Roles and responsibilities should be clearly defined within this organization. A number of *system improvement teams* should be formed, either permanently or on an on-demand basis. Such a cross-functional team should be led by a process owner who is responsible for the design, implementation, operation and performance improvement of that particular system process/function.
- *Competent procedures and tools*. Ideally, an information environment should be set up to formalize MSM procedures and to assist in the execution of an MSM cycle. It should help capture and document strategic data and MSD decisions, and should provide training materials when necessary. For example, the task-centered way in which this handbook is structured and presented makes the tasks ideally suited for adaptation on a company's intranet-based information system. The generic structure and functionality of the TCMM presented in Section 4.6 provides one of the possible platforms for this purpose, as illustrated in Figure 7.1.

The following cases illustrate respectively the organizational structure and the information environment to facilitate MSM's application and institutionalization in practice.

TASK-CENTRED MSM WORKBOOK
SYSTEM OPERATIONS MANUALS

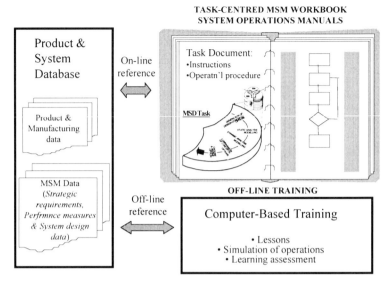

Figure 7.1 MSM procedures within a task-centered information environment

7.2 CASE A: MSM ENABLED ORGANIZATION

The background of this case was presented in Chapter 1. The following provides additional material to illustrate how the MSM is institutionalized within the organization according to the three key requirements: people, structure and information environment.

7.2.1 MSD Procedures

As part of the institutionalization of MSM, the company developed a particular MSD task procedure called business process design (BPD). This was populated within the MSD area in order to design all the processes involved in the greenfield MSD project. Figure 7.2 shows the logical position of this MSD process and its links with other aspects in the MSM framework:

- *Point 1*: link to business strategy, strategic and customer focus of process, and initiation and approval of process design and change.
- *Point 2*: link to organization's management system, documentation of changes, input from process reviews and audits, advanced process audits, accreditation compliance check during process design.
- *Point 3*: alignment of organization according to process changes, design of new jobs, competence profiling, recruitment (internal and/or external) to fill new positions, training of all affected people, changes to working structures and patterns, changes to remuneration, pay grading.
- *Point 4*: planning, design and implementation of IT applications, implementation of IT infrastructure and hardware.

- *Point 5*: specification and procurement of equipment required in new process.
- *Point 6*: design and implementation of layout (office and shopfloor space, etc.)

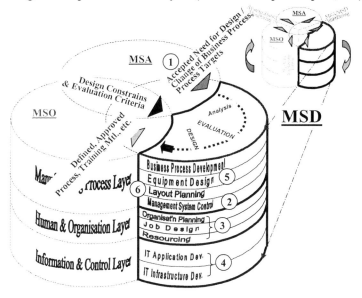

Figure 7.2 BPD integrated within the MSM framework

All these interfaces need to be managed when designing a new process or changing an existing one. This MSD procedure and the interfaces are fully established in the organization. It can be initiated and the associated tasks followed through at any time, so that there is no need for the company to recreate this process each time an MSD project takes place. This has helped the company personnel involved in MSD activities to concentrate on the actual design tasks, rather than having to create and establish a design methodology first, and then sell it to the management of the organization.

7.2.2 Structure of SIT Teams

The company established SIT teams within its organizational structure to facilitate the following activities:

- Process design/change authorization: deciding and reviewing the need for change, and authorizing the project.
- Process owner identification: choice of which department/person will own a process.
- Identification and adaptation of relevant design tools: worksheets such as analysis matrix, checklist, and formation of MSD project plans.

The case experience has shown that the above represented some of the most difficult aspects because there was a high degree of uncertainty involved, and a high potential for political issues to surface. It was therefore of great importance to follow the principles of change and human resources management described in this workbook. An SIT team in this case consisted of:

- *Function owner*—the leader of the team who owns one or more functions and is responsible for their design, implementation and operation. Future developments and improvements are also to be driven by the function owner, who is responsible for the overall performance of the function(s) concerned.
- *BPD coordinator*—a design expert for a particular function, and a representative in the BPD steering team, responsible for the coordination of the design activities, and integration and coaching of function owners.
- *BPD steerer*—who has the overall control of the BPD process, including the design and implementation of MS functions, and authorization of new projects.
- *Function designers*—involving cross-functional personnel, such as customers, suppliers and contributors of the function to be designed. These people are responsible for its analysis, evaluation, design and implementation.

7.2.3 MSD Project Management

The parallel design and implementation of around 70 MS functions was a challenge for the organization as a whole. A capable project management process was required to ensure the on-time design, implementation and integration of all functions concerned. For this reason a significant element of the BPD process was about project management. The key tools adapted by the company included (Chapter 1):

- *MS project checklist*—a checklist of all the functions to be designed (equivalent to *Worksheets MSA/MSD 2.1.1*: MSD Project Formulation, and *Worksheet MSA/MSD 2.2.1*: Terms of Reference).
- *MS analysis and design matrix*—planning and monitoring of the analysis and design of individual functions (equivalent to *Worksheets MSA/MSD 2.3.1*: MSD Task Selection, and *MSA/MSD 2.4.1*: Project Execution Plan).
- *MS implementation checklist*—project management of implementation and operation of individual functions (equivalent to *Worksheet MSI 2.1*: Implementation Plan).

For each of the SIT team members, the project checklist stated the high level activities that should be carried out within the three MS layers (*processes, IT, human and organization*) through each of the MSM phases. The team had to decide when it would carry out these activities, working backwards from the "go-live" date of the process. This approach could also lead to phased implementations, where a series of go-live dates are used to satisfy the need to implement certain parts of a function earlier than others. Each of the phases had a gateway at which the teams held formal reviews and reported the latest project status. The most prominent milestone was, of course, the "go-live" date of the entire process, representing the maturity of the function.

Process	Process	Process Owner	Date	Status	Date	Status	Date	Status	Date	Status	Comments
	Quality										
QMP 05.01	The Writing and Change of Systems Documents	D. Mills	Jul-99		Aug-99		Sep-99		Jan-00		
QMP 05.02	Systems Audits	D. Mills	Jun-99		Aug-99		Sep-99		May-00		
QMH	Quality Management Handbook	C. Davies	Aug-99		Sep-99		Oct-99		Jan-00		
QMP 04.01	Devt & Cascade of Business Objectives & Targets	C. Davies	Jul-99		Sep-99		Oct-99		Jan-00		Proc comp, reqs prestn to 1st line & 2nd linecascade (07/00)
QMP 05.06	Roles & Responsibilities regarding Quality Issues	C. Davies	Sep-99		Oct-99		Oct-99		Jan-00		
QMP 17.01	Document and quality records	D. Mills	Aug-99		Sep-99		Oct-99		Jan-00		
QMP 12.01	Measuring Room	P. Ward	Sep-99		Nov-99		Nov-99		Feb-00		
QMP 12.02	Laboratory	P. Avis	Sep-99		Nov-99		Nov-99		Feb-00		
QMP 13.03	Gauge Calibration & Control	P. Ward	Sep-99		Sep-99		Nov-99		Feb-00		
QMP 05.04	Dynamic Test	J. Fletcher	Jul-99		Sep-99		Dec-99		Apr-00		
QMP 05.05	Static Audit	J. Fletcher	Jul-99		Oct-99		Dec-99		Apr-00		
QMP 09.02	Supplied Parts Quality Activities	P. Collins	Jan-98		Jun-99		Jan-00		Apr-00		Issue 2 under development for new responsibilities
QMP 14.01	Management of Non-Conforming Material	D. Lucas	Jun-99		Nov-99		Mar-00		Sep-00		
QMP 15.01	Problem Resolution	D. Burt	Oct-99		Mar-00		Apr-00		Dec-00		
QMP 11.13	Product Traceability	D. Burt	Apr-00		Jul-00		Sep-00		Dec-00		
QMP 16.02	Quality Control	D. Burt	Dec-99		Aug-00		Nov-00		Feb-01		Model agreed, C/Plant Q2, QBM & Warranty SOP
QMP 06.01	Periodic and Yearly Quality Reports	C. Davies	Nov-99		Sep-00		Dec-00		Mar-01		
	Finance										
GMP 04.01	Customer Processing	T. Summerer	Dec-98		Jul-99		Jan-00		Jul-00		Sub-proc. to "go live" in later releases, dates tbd.
GMP 04.02	Supplier Processing	N. Flaherty	Dec-98		Jul-99		Jan-00		Jul-00		Being redefined to include new Purchasing processes, likely completion towards year end, training extending
GMP 04.03	Investment Processing	S. Connellan	Dec-98		Jul-99		Jan-00		Jul-00		Process redefined undergoing approval sign off.
GMP 04.04	Financial Accounting	T. Summerer	Dec-98		Jul-99		Jan-00		Jul-00		Sub-proc. to "go live" in later releases, dates tbd.
GMP 04.05	Controlling	R. Cooke	Dec-98		Jul-99		Jan-00		Jul-00		(as above). Roll-out in line with SAP 4.6 Upgrade
GMP 04.06	Customs Clearance	H. Seidl	N/A				Dec-00		Jan-01		Undergoing Sign-off
GMP 04.08	Technical Purchasing	B. Weiser	Nov-00		Jan-01		Feb-01		May-01		Training of Purchasing Team sufficiently progressed. Link to AG processes
GMP 04.07	Manpower Planning	S. Connellan	Nov-00		Nov-00		Mar-01		Jun-01		Process designed and being used. Documentation under development, dependant on SM-65 GMP's

Figure 7.3 Project status report of the case company

The status of each design project—including the status of the function checklist shown in Figure 7.3—was reported regularly to the top management of the company. Due to its successful application on the greenfield project, the company's MSM setup (i.e., the structure and procedures outlined above) are now formally incorporated into the company's overall business management system to enable the factory's future system improvement (Figure 7.4).

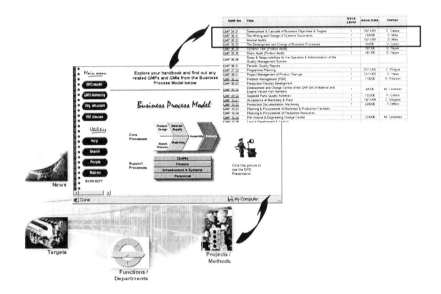

Figure 7.4 Institutionalization of MSM procedures as part of the organization's management system

7.3 CASE B: ONLINE OPERATION PROCEDURES AND TRAINING

TCMM is considered an enabler for the institutionalization of MSM procedures within an organization. This case provides a more detailed account of its structure and operation. Although not specifically related to a system design project, it illustrates some key features of such an information platform, such as online operations documentation and on-demand training. These features can be used to provide support for both the normal system operations and the MSM framework and its task documents.

In this case, the traditional ways of providing manufacturing information to the shop floor were not necessarily task-related. Rather, general information was available but needed to be found when required. Although this satisfied the requirements of normal operations, there were a number of problems associated with this form of documentation. Some of the key problems included: the physical separation of the processes, their descriptions and procedures; the poor user friendliness; the high maintenance efforts; and the inability of the documentation systems to effectively capture process "know how". In contrast, a task-centered, multi-media MS information system utilized a web-based database of reference manuals to provide company personnel with comprehensive tools for looking up procedures and product information. It can support multimedia objects and has the added benefits that both authoring and viewing tools are widely used and well known. Furthermore, the approach was cost-effective, easy to install and highly flexible. It allows for change without major systems development efforts, and the skill requirements are relatively low. The HTML (Hyper Text Markup Language) front-end can be connected to a database back-end if required. Additional features include:

- *Adobe PDF format.* As a widely accepted standard, PDF (Portable Document Format) is perhaps the most suitable format for electronic documentation for this kind of application. PDF files are compatible with HTML files, essentially platform independent, and communicate well with any web server or browser.
- *Search engines and automatic indexing.* A search engine speeds up the process of finding the required topic in the system.

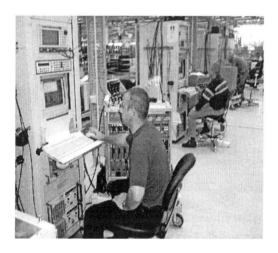

Figure 7.5 Testing of transceiver stations

This case provides an example of actual TCMM application. The collaborating company involved in this case is part of a global provider of integrated communications solutions. The organization is dedicated to the research, development and manufacture of GSM (Global System for Mobile Communications) equipment, the digital standard adapted worldwide for mobile telephone technology. The manufacturing processes involve the production and assembly of mainly base transceiver stations. These are used as part of the infrastructure to support the providers of GSM services. Worldwide demand for the equipment is such that the manufacturing process is continuous 24 hours a day, seven days a week. The environment in the assembly area of the base transceiver stations is highly automated but the human factor was important in testing the final products (Figure 7.5). It remained labor intensive, and depended on the operators' experience. With many varieties of configurations, the traditional approach made it difficult to guarantee the standards and quality of the operations.

The company attempted to consolidate its manufacturing processes by maximizing the use of its resources in personnel, and information technology. In particular, the organization was developing a generic platform to enhance the efficiency of its production test facilities. Its objective was to provide an infrastructure for communication, sharing and recycling resources, reducing test development cycle time, minimizing manual operations, and improving the fault-finding processes. Within this environment, all the engineers shared their experience in various aspects of systems engineering and developed test systems concurrently. The system also aimed to provide a series of tools and functions to be used throughout the factory for the testing of multiple products. As part of the company's overall initiative, a fully functional TCMM system was developed, providing a working environment to train the company's new operators, as well as its joint venture partners in different parts of the world. Utilizing the TCMM concept, the two main objectives identified were to develop a system that:

- Supplies the testing area personnel with a comprehensive tool for looking up technical information about products, testing equipment and procedures. This should be of use to first time operators, as well as skilled technical personnel.
- Provides a tool that can teach a first time operator to test a product from start to finish with either minimal or no external training. The system should also provide an assessment tool for the qualification of the trainees, and for recording the performance of skilled personnel.

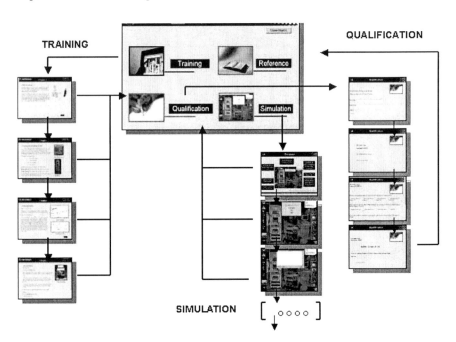

Figure 7.6 Structure of the TCMM system

The overall system structure is as shown in Figure 7.6, with the following modules.

- *Reference module.* The reference module serves as the knowledge repository of the system. It contains all the technical information relevant to the testing procedures, the base stations, and the technology used. The reference module can be readily accessed within the system whenever the users require more in-depth technical information about a subject. The information contained in the reference module is organized into four related parts. Each of these parts is then subdivided into smaller sections in order to make the retrieval of information faster and easier. All these parts are linked, either directly or indirectly, to help in cross-referencing. Organizing the information in this way also facilitates the maintenance/updating of data.
- *Training module.* The training module provides a new trainee with introductory information about four different subjects: an overview of GSM and its component parts, product information, equipment information (cable connections required between different instruments in order to setup tests), and test information (procedures that an operator must follow in order to test a product). The subjects are presented in sequential chapters. This format was considered the most appropriate for training purposes since the trainee is required to cover all the material included in the desired sequence. At the end of each chapter, the trainee has the option to carry out a self-assessment. This facility provides him/her with feedback on the progress made.
- *Simulation module.* The simulation module is a subset of the training module. It provides a virtual environment of the testing area, and a suite of tools that allows a trainee to learn and try out a complete cycle of the testing process (Figure 7.7). The system is interactive with the trainee throughout the simulation run. It provides step-by-step instructions, a list of options for each action to be carried out, and possible tools and devices. Icons symbolizing the tools, devices and plugs needed during the testing process are available in the right column, and can be clicked if the trainee requires a particular item during the exercise. The system then monitors the actions undertaken by the trainee and, depending on whether the required one is selected, either continues the operation or offers further assistance. At any time during the simulation cycle, the trainee has access to all the product information and operational documentation. This online facility is useful for finding answers to questions that the trainee may have regarding an operation. In addition, video clips are available to provide further guidance. The simulation process was developed mainly using Dynamic HTML, which allowed the development of a virtual environment for interactive actions that can be performed during training.
- *Assessment module.* This completes the logical cycle of training that is supported within the TCMM environment (lessons/simulated-operation/qualification). The qualification module developed follows a straightforward procedure. To start the assessment, a set of questions is selected randomly from a database. The trainee's choice of answer is assessed, and results are recorded for both self-assessment and employee qualification.

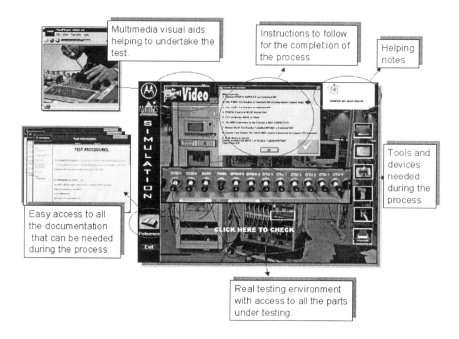

Figure 7.7 Virtual simulation environment

The management of the company carried out a detailed survey to evaluate the effectiveness of the system. Feedback from these was very positive. It was pointed out that, compared with the existing approaches that leave the users almost entirely on their own to identify relevant data/information to support the manager/operators' current work, the TCMM working environment equips the user with a structured, user-friendly way to make use of company information/operational manual/data. In general, the concept of the TCMM approach provides a logical implementation foundation, providing a general mechanism for task/tool/data integration, so that the operator/system design is given direct, structured and ready access to relevant information and tools. Its practical applications to date have illustrated clearly its value both as a self-contained information system, and as a supplementary system to the existing databases and other information applications.

Furthermore, its structure as a knowledge repository can easily adapt as the company's product ranges, MS processes and MS system structure progress through time.

Selected Bibliography

Berry, W.L., and Hill, T.J., 1992, Linking systems to strategy, *International Journal of Operations and Production Management*, **12**, 10, pp 3-15.

Black, J.T., 1991, *The Design of the Factory with a Future* (McGraw-Hill).

Carrie, A.S., *et al.*, 1994, Linking strategy to production management structures and systems, *International Journal of Production Economics,* **34**, pp 293-304.

Doumeingts G., Vallespir, B., Chen, D., 1995, Methodologies for designing CIM systems: a survey, *Computers In Industry*, **25**, pp 263-280.

GAO (USA General Accounting Office), 1998, *Performance Measurement and Evaluation: Definition and Relation*, GAO/GGD-98-26, USA, General Accounting Office.

Hax, A.C., and Majluf, N.S., 1991, *The Strategy Concept and Process, a Pragmatic Approach* (Prentice Hall, London).

Hayes, R.H., and Pisano, G.P., 1994, Beyond world class: the new manufacturing strategy, *Harvard Business Review*, **72**, pp 77-86.

Hill, T., 1995, Manufacturing Strategy, Text and Cases (Macmillan, Basingstoke).

Kasul, R.A., and Motwani, J.G., 1995, Performance measurements in world-class operations: a strategic model, *Benchmarking for Quality Management & Technology*, **2**, 2, pp 20-36.

Kim, J.S. and Arnold, P., 1996, Operationalizing manufacturing strategy, an exploratory study of constructs and linkage, *International Journal of Operations and Production Management,* **16**, pp 45.

Kleinhans, S., Merle, C., Doumeingts, G., 1995, Determination of what to benchmark: a customer-oriented methodology, *Benchmarking—Theory and Practice,* edited by Rolstadas, A., pp 267-276 (Chapman & Hall, London).

Leong, G.K., and Ward, P.T., 1995, The Six Ps of manufacturing strategy, *International Journal of Operations and Production Management*, **15**, 12, pp 32-45.

Parnaby, J., 1986, The design of competitive manufacturing systems, *International Journal of Technology Management*, **1**, 3/4, pp 385-396.

Skinner, W., 1985, *Manufacturing: The Formidable Competitive Weapon* (John Wiley & Sons, Chichester).

Tranfield, D., Smith, S., 1990, *Managing Change, Creating Competitive Edge* (IFS Publishing, Kempston).

Troxler, J.W., and Blank, L., 1989, A Comprehensive Methodology for Manufacturing System Evaluation and Comparison, *Journal of Manufacturing Systems,* **8**, 1, 175-183.

Wiendahl, H.P., and Scholtissek, P., 1994, Management and control of complexity in manufacturing, *Annals of the CIRP*, **43**, 2, pp 533-540.

Wu, B, 1994, *Manufacturing System design and Analysis*, 2nd Edition (London: Chapman & Hall).

Wu, B., 1996, An overview of the technical requirements for an integrated computer-aided manufacturing system design environment, *International Journal of Manufacturing System design*, **2**, 1, pp 61-72.

Wu, B., 1997, Integrated CAMSD, *Proceedings of the 32nd MATADOR* (UK: University of Manchester Institute of Science and Technology), pp 41-47.

Wu, B., 2000, *Manufacturing and Supply Systems Management: A Unified Framework of System design And Operation* (Springer-Verlag, London).

Wu, B., *et al*, 2000, The design of business processes within manufacturing systems management, *International Journal of Production Research*, **38**, 17, pp 4097-4111.

Wu, B., 2000, Manufacturing strategy analysis and system design: the complete cycle within a computer-aided design environment, *IEEE Transactions on Robotics and Automation,* **16**, 3, pp. 247-258.

List of Task Documents and Worksheets

Task Document MSM—MS Status Monitoring 263

Index